THE INVENTION OF HANS COLER

RELATING TO AN ALLEGED NEW SOURCE OF POWER

by R. Hurst

British Intelligence Objectives Sub-Committee

1946

BIOS Target Number C31/4799

Unclassified from CONFIDENTIAL status

Integrity Research Institute
5020 Sunnyside Avenue
Beltsville MD 20705

B.I.O.S. FINAL REPORT No. 1043.

COPY No. 306

ITEM No. 31.

~~CONFIDENTIAL~~

UNCLASSIFIED IN ACCORDANCE WITH FILE [illegible]

Min. Tech. TIL (G)

THE INVENTION OF HANS COLER, RELATING TO AN ALLEGED NEW SOURCE OF POWER.

This report is issued with the warning that, if the subject matter should be protected by British Patents or Patent applications, this publication cannot be held to give any protection against action for infringement.

~~CONFIDENTIAL~~

BRITISH INTELLIGENCE OBJECTIVES SUB-COMMITTEE

Released in the Summer 1946

THE INVENTION OF HANS COLER

RELATING TO AN ALLEGED NEW SOURCE OF POWER

by R. Hurst

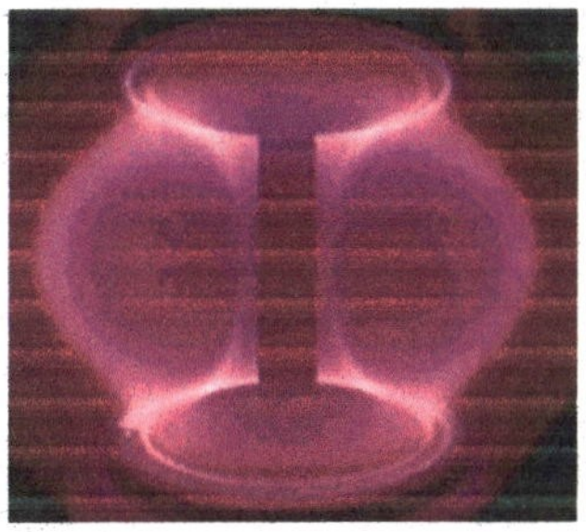

British Intelligence Objectives Sub-Committee

1946

BIOS Target Number C31/4799

Unclassified from CONFIDENTIAL status

Integrity Research Institute
5020 Sunnyside Avenue
Beltsville MD 20705

CONFIDENTIAL

THE INVENTION OF HANS COLER, RELATING TO AN ALLEGED NEW SOURCE OF POWER

Reported by R. Hurst, M. of S.

B.I.O.S.Trip No. 2394

BIOS Target Number: C31/4799.

BRITISH INTELLIGENCE OBJECTIVES SUB-COMMITTEE,
32, Bryanston Square, London, W.1.

CONFIDENTIAL

TABLE OF CONTENTS

Personnel of Team

R. Hurst, Ministry of Supply
Captain R.Sandberg, Norwegian Army

The Invention of Hans Coler relating to an alleged new source of power

I. OBJECT OF VISIT AND SUMMARY

Coler is the inventor of two devices by which it is alleged electrical energy may be derived without a chemical or mechanical source of power. Since an official interest was taken in his inventions by the German Admiralty it was felt that investigation was warranted, although normally it would be considered that such a claim could only be fraudulent.

Accordingly Coler was visited and interrogated. He proved to be co-operative and willing to disclose all details of his devices, and consented to build up and put into operation a small model of the so-called "Magnetstromapparat" using material supplied to him by us, and working only in our presence. With this device, consisting only of permanent magnets, copper coils, and condensers in a static arrangement he showed that he could obtain a tension of 450 millivolts for a period of some hours: and in a repetition of the experiment the next day 60 millivolts was recorded for a short period. The apparatus has been brought back and is now being further investigated.

Coler also discussed another device called the "Stromerzeuger", from which he claimed that with an input of a few watts from a dry battery an output of 6 kilowatts could be obtained indefinitely. No example of this apparatus exists today, but Coler expressed his willingness to construct it, given the materials, the time required being about three weeks.

Opportunity was taken to interrogate Dr. F. Modersohn who had been associated with Coler for ten years and had provided financial backing. He corroborated Coler's story in every detail.

Neither Coler nor Modersohn were able to give any theory to account for the working of these devices, using acceptable scientific notions.

II HISTORICAL NOTES

1. The "Magnetstromapparat"

This device consists of six permanent magnets wound

in a special way so that the circuit includes the magnet itself as well as the winding, (See Fig.1). These six magnet-coils are arranged in a hexagon and connected as shown in the diagram (Figs. 2 and 3), in a circuit which includes two small condensers, a switch, and a pair of solenoidal coils, one sliding inside the other. To bring the device into operation the switch is left open, the magnets are moved slightly apart, and the sliding coil set into various positions, with a wait of several minutes between adjustments. The magnets are then separated still further, and the coils moved again. This process is repeated until at a critical separation of the magnets an indication appears on the voltmeter. The switch is now closed, and the procedure continued more slowly. The tension then builds up gradually to a maximum, and should then remain indefinitely. The greatest tension obtained was stated to be 12 volts.

The "Magnetstromapparat" was developed by Coler and von Unruh (now dead) early in 1933, and they were later assisted by Franz Haid of Siemens-Schukert, who built himself a model which worked in December 1933. This was seen by Dr. Kurt Mie of Berlin Technische Hochschule and Herr Fehr (Haber's assistant at the K.W.I.), who reported that the device apparently worked, and that they could detect no fraud. One model is said to have worked for 3 months locked in a room in the Norwegian Legation in Berlin in 1933. No further work appears to have been done on this system since that date.

2. The "Stromerzeuger"

This device consists of an arrangement of magnets, flat coils, and copper plates, with a primary circuit energised by a small dry battery. The output from the secondary was used to light a bank of lamps and was claimed to be many times the original input, and to continue indefinitely. Details of the circuit, and a theory as to its mode of operation were given (summarized in Appendix I).

In 1925 Coler showed a small (10-watt) version to Prof. Kloss (Berlin), who asked the Government to give it a thorough investigation, but this was refused, as was also a patent, on the grounds that it was a "perpetual motion machine". This version was also inventions, when the details above were given.

He was then questioned about the theory of the devices, but he was unable to give any coherent suggestions as to the mechanism. He stated that his researches (apparently conducted with crude apparatus) into the nature of magnetism had lead him to conclude that ferro-magnetism was an oscillating phenomenon, of frequency about 180 kilohertz. This oscillation took place in the magnetic circuit of the apparatus, and induced in the electrical circuit oscillations the frequency which of course depended on the values of the components used. These two phenomena interacted, and gradually built up the tension. As the mechanism was not understood the proper arrangement could not be worked out, but had been arrived at by experiment, and the apparatus had to be brought into adjustment by similar trial and error methods. Coler stated that the strength of the magnets did not decrease during use of the apparatus; and suggested that he was tapping a new sort of energy hitherto unknown, - "Raumenergie" (Space-energy). Coler gave a resume of the work done by Dr. Frohlich for the O.K.M., and produced a copy of Frohlich's report, written jointly with himself (translation reproduced as Appendix IV) and a report of his own (part of which is given in Appendix V).

Coler was next asked if he would consent to build models of these devices if material was made available. He agreed that he could do this, and stated that it would take one week to construct a "Magnetstromapparat", and a month to construct a "Stromerzeuger". Accordingly we supplied the magnets, condensers and copper wire needed for the former, and Coler proceeded to build an apparatus as discussed in Section 3. A list of material required for the "Stromerzeuger" was drawn up by Coler.

2. Interrogation of Dr. F. Modersohn

Modersohn was questioned about the history of these devices, with which he had been concerned financially, and corroborated the details given by Coler He stated that he had at first disbelieved Coler's claim but had taken great precautions to eliminate fraud. He had seen the 70 watt "Stromerzeuger" working on a number of occasions, and had taken it while working from one room to another. All parts were visible and nothing was hidden. As he was himself not expert he had tried to get experts to examine it thoroughly, but reputable scientists either refused to have anything to do with it at all, or else were more concerned to find a fraud than to see how it worked. The exception was Dr. Frohlich, who was convinced of the reality of the effect, and who also believed that the secret was to be found in an analysis of the energy changes in the special inductive circuit used. He had made experiments to test his ideas, but Modersohn denied knowledge of the results. Modersohn was extremely methodical, and showed his files on the subject: these contained copies of all letters and reports concerning the device, since 1933.

3. Construction and Testing of the "Magnetstromapparat"

In our presence and with material supplied by us (some brought from England and the rest bought locally) Coler built an apparatus as shown in Figs.1, 2, and 3. It is to be noted that some magnets are wound in a clockwise direction looking at the N pole (called left) and others in an anti-clockwise direction (called right). The magnets were selected to be as nearly equal in strength as possible, and the resistance of the magnet-coil combination was checked after winding to see that this also was uniform (about .33 ohm). The physical arrangement was as shown in Fig. 2, in a breadboard style. Measurements of voltage and current across A-B were made with a Mavometer. A mechanical arrangement of sliders and cranks for separating the magnets evenly all round was made up.

On 1.7.46 experiments were being continued after three days of fruitless adjusting, and when the magnets were at a separation of about 7 mm. the first small deflection was noted (about 9 a.m.). The switch was closed and by slow adjustment of the sliding coil, and by increasing the separation of the magnets to just over 8 mm, by 11 a.m. the tension was raised to 250 millivolts and by 12.30 p.m. it was 450 millivolts. This was maintained for another 3 hours when a soldered tag became disconnected, and the meter slowly dropped back to zero. Soldering up the broken connection did not restore the tension. The magnets were closed up and left overnight and the same procedure for finding the adjustment was repeated on 2.7.46. After about three hours a deflection of 60 millivolts was obtained; this was maintained for more than 30 minutes, but then decreased to zero when further adjustments were tried.

During all this work the model was completely open, and nothing could be hidden in it. The breadboard and meter could be picked up and moved round the room, tilted, or turned, without effect.

The apparatus would appear to be too crude to act as as a receiver of broadcast energy, or to operate by induction from the mains (the nearest cable being at least 6 feet away), and the result must for the moment be regarded as inexplicable.

IV. CONCLUSIONS

1. It was judged that Coler was an honest experimenter and not a fraud, and due respect must be paid to the judgment of Frohlich in the matter as deduced from his report to Seysen.

2. The result obtained was genuine in so far as could be tested with the facilities available, but no attempt has yet been made to find an explanation of the phenomenon.

3. It is felt that further investigation by an expert in electromagnetic theory is warranted, and that Coler's offer to construct a model of the "Stromerzeuger" should be taken up.

APPENDIX I

Summary account of the "Stromerzeuger"

1. The basic principle is that an electron is to be regarded not only as a negatively charged particle but also as a South magnetic pole.

The basic element is that of an open secondary circuit, capacity loaded, inductively coupled to a primary circuit. The novel feature is that the capacities are connected to the secondary core <u>through permanent magnets,</u> as shown:-

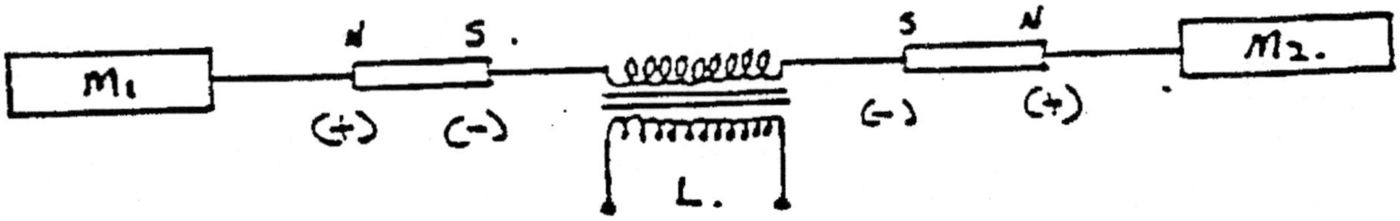

It is claimed that, on switching on the primary circuit, "separation of charges" takes place, i.e. M_1 becomes + vely charged and M_2 - vely charged, and that these charges are "magnetically polarised" when they are formed, owing to the presence of the magnets. On switching off the primary circuit a "reversing current" flows in the secondary, but the magnets "do not exert a polarising effect on this reversal".

2. Two of these basic elements are now placed together making a double system or stage with the plates close together in parallel planes as shown:-

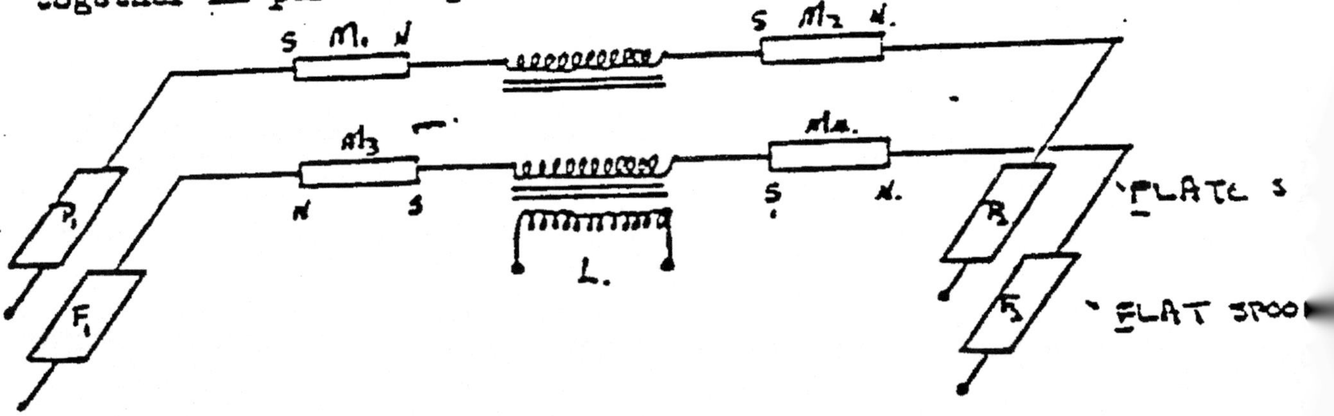

The secondary windings are both exactly equal and wound in a direction such that on switching on the primary coil the electrons in the secondary coil flow from P_1 to P_2 and F_1 to F_2.

It is then stated that system $F_1 - F_2$ merely has an inducing effect, and the useful current comes into existence in the system $P_1 - P_2$.

3. A single stage cannot be effective but two stages connected so that the numbers of effective North and South poles are equal will provide a basic working arrangement. More double stages can then be added to provide higher outputs.

4. It is then stated that as well as the normal electrons flowing from the battery and from induction when the circuit is opened or closed "space electrons" flow from "repelling spaces" to "attracting spaces" between the plates, but this theory I was unable to follow any further.

APPENDIX II

Translation of Report by Professor M. Kloss (Berlin) on tests on the apparatus, constructed by Captain Coler (dated 4th March, 1926)

On the 3rd of March I have on request inspected the apparatus constructed by Captain Coler, and I undertook some tests regarding its efficiency. Professor Dr. R. Franke from the Technical College of Berlin was present at these tests.

The apparatus, in its construction already known to me from a previous inspection, consists of:

a double-row system of copper-plates,
a double-row system of flat spools, and
a system of electro-magnets,

to whose cores silver wires are attached and through which branch currents of the plate-system are conducted. For each of the three systems a three part accumulator-battery of 6 volts and 6.5 ampere hours capacity is provided. The plate-circuit and the spool-circuit are parallel so that the two batteries appertaining thereto can also be replaced by a single battery. This was ascertained towards the end of the tests by switching off the one battery while the apparatus was working.

When asked why two batteries were used at all, Captain Coler declared that for starting the apparatus a double battery is necessary to get a second charge-impetus after exciting with the one battery, and this for releasing the peculiar character of the apparatus. A test made towards the finish of a series of trials confirmed this assertion in as much as the mechanism could not be started with the single battery, on the contrary the "adjustment" of the mechanism got disturbed.

Current-indicators are built into each of the three circuits mentioned, as well as volt-meters, behind some switch-resistances necessary for the "adjustment". Between the open ends of the two plate and spool-systems there are the terminal clamps for the effective circuit for whose loading three bulbs of 8 volts are provided. The apparatus was then put into action and above all the load was tested with the aid of the built-in instruments, that is on being loaded with 2 lamps, 3 lamps and running light. Thereupon and by means of precision-instrument (continuous current type) Siemens & Halske Nr. 423820 the indications of the built-in instruments were controlled by switching on the mentioned precision-instrument in turn to the single circuits, that is direct to the one terminal clamp of the battery, in order to ascertain if perhaps through any by-connections with the batteries, not conspicuous at once, more current might be received than the built-in instruments were showing. The tests showed a sufficient conformity within the bounds of measuring accuracy.

Finally the same instrument was switched on also to the effect circuit, whereby some greater differences with the instruments built into this circuit were shown; on the other hand, however, not such differences as to alter fundamentally the total result in any way.

In order to find out besides, whether the output current might be identical with a wave-current overlapping continuous current, the last control test was repeated by using a hotwire-instrument furnished by me, make of Hartmann & Braun, Nr. 254159. In case there would have been such wave-currents, the hotwire-instrument would have had to indicate a bigger current than was shown by the continuous-current instrument. In reality, however, a somewhat smaller current was found when employing the hotwire-instrument. This may be explained by the fact that the inner resistance of the instrument is about 10 times greater than that of the continuous-current instrument, so that when switching on the hotwire-instrument the load of the apparatus was somewhat lessened. The difference between 3.08 ampere measured with the continuous-current instrument as compared with 2.95 ampere measured with the hotwire-instrument is easily explained when taking into account the divergences of the two instrument-resistances. Therefore one can infer from this test that in the load-circuit we have to deal with real continuous current.

The results of the tests are compiled in the annexed table.

The figures show very well that the consumption of energy in the external circuit is greater than the energy taken from the batteries. According to the circuit, produced by Captain Coler, which within this short time I could not check in all its parts, the magnet-exciting circuit is fed by a special battery, completely separated from the other two circuits. Consequently, a direct comparison of efficiency and consumption of the apparatus would mean that only the sum of current of the plate circuit and of the spool circuit would count. After the established estimates with my own instrument and on a load of three bulbs, there was resulting a current from the two mentioned batteries of 0.215 + 0.070 = 0.285 ampere. At the same time the three bulbs consumed ca. 3.7 ampere, according to the built-in instruments, which is about 0.2 ampere too much as was proved later on by a control of this instrument, so that the real consumption has been about 3.5 ampere at a tension of about 2.3 volt.

The reception of current from the two batteries in this case consequently was 1.7 watt while the consumption of the seen by Profs. Schumann (Munich), Bragstad(Trondheim) and Knudsen (Copenhagen). Reports by Kloss and Schumann are translated in Appendices II and III.

In 1933 Coler and von Unruh made up a slightly larger model with an output of 70 watts. This was demonstrated to Dr. F. Modersohn, who obtained from Schumann and Kloss confirmation of their tests in 1926. Modersohn then consented to back the invention, and formed a company (Coler G.m.b.h.) to continue the development. At the same time a Norwegian group had been giving financial support to Coler, and these two groups clashed. Modersohn's connection with Rheinmetall Borsig, and hence with the official Hermann Goering combine gave him an advantage in this. Coler then in 1937 built for the Company a larger version with an output of six kilowatts.

In 1943 Modersohn brought the device to the attention of the Research Department of the O.K.M. The investigation was placed under the direction of Oberbaurat Seysen, who sent Dr. H. Frolich to work with Coler from 1.4.43 to 25.9.43. Frohlich was convinced of the reality of the phenomena, and set about investigating the fundamentals of the device. He apparently concentrated on a study of the energy changes which occur on the opening and closing of inductive circuits. At the end of the period he was transferred to B.M.W. to work on aerodynamic problems and is now working in Moscow.

In 1944 a contract was arranged by the O.K.M. with Continental Metall A.G. for further development, but this was never carried out owing to the state of the country. In 1945 the apparatus was destroyed by a bomb, in Kolberg, whither Coler had evacuated. Since that time Coler had been employed, sometimes as an engineer and sometimes as a labourer. Modersohn had severed his connection with Rheinmetall Borsig, of which he had been a director, and was working for the Russian authorities as a consultant in chemical engineering.

III REPORT

1. Interrogation of Coler

Coler was questioned first about the history of his

bulbs amounted to about 8 watt. Especially striking in this connection is the considerably higher current-power in the bulb-circuit being about 12 times bigger than the current coming from the two batteries.

We have also absolutely made sure that from the batteries no other conductors led to the apparatus than those into which my instrument was built-in. The fact that an increase of power from the battery to the terminal clamps of the effective circuit in the plate-system takes place, could, indeed, not be tested on the different parts of the apparatus by a direct measuring of the power itself, because Mr. Coler declared that when switching on an instrument in the interior of the system, probably the "adjustment" would be disturbed.

I have therefore tested the decrease of tension in the single plates on a load of three lamps by means of a millivolt-meter, make of Hartmann & Braun, Nr. 462375, in order to get at least in an indirect way an explanation for the increase of current. This examination showed a remarkable increase of tension-losses with a distinct maximum on the third-last plate of the one row. These estimates too are compiled in the table at the end of this judgment.

Results

The result of the investigation showed an astonishing working of the apparatus, which, without further researches cannot be explained or compared with the hitherto known characteristics.

Regarding the Cause of the observed characteristics, especially the solution of the question, where the energy in the apparatus originates, no explanation can be given yet, after the short and simple tests.

Solely the conjecture can be expressed that the magnet-system is the source of the energy. It therefore seems desirable to clear this point by further thorough and systematic examinations. For this purpose the proposition of Mr. Coler to construct a bigger apparatus is very sensible, an apparatus, into which from the beginning measuring-instruments should be built in the respective parts of the single circuits in order to enable the "adjustment" of the apparatus including these instruments.

A judgment for the possible utilization of the phenomena,

observant in the apparatus, from the economical point of view, I am, of course, not able to pass on after these short tests. This question could only be answered after some lengthened and permanent tests, by which should be established what productiveness the source of energy, existent in the apparatus will show, especially if the magnets, in case they are the supports of the energy-source would wear out after some time, and at what expense each time they could be replaced by new ones.

Above all, however, it appears important that the part or parts of the apparatus, being possibly responsible for causing the observed phenomena, should be taken out in a way as simply as possible and be submitted to an examination in all directions.

Before finishing this, I wish to say that the results of the tests are put at the disposal of Captain Coler for which reason I enclose a copy herewith.

I should like to ask, however, the gentlemen in question not to mention my name and that of Professor Franke nor divulge the results of our tests without our express consent, or to make them known publicly and above all not in the press.

(Signed) Dr. Ing. M. Kloss
Professor

Results of Measuring

Apparatus Coler
3rd March, 1926

Nr.	Test with instrument	Charge	Magnet circuit		Plate circuit		Spool circuit		External circuit	
			Amp.	Volt	Amp.	Volt	Amp.	Volt	Amp.	Volt
1	built-in	-	0.21	0.5	-	6.4	-	6.3	-	6.0
2	idem	2 bulbs	0.21	0.5	0.16	4.7	0.06	4.3	3.1	3.5
3	idem	3 bulbs	0.21	0.5	0.22	4.0	0.075	3.45	3.7	3.0
	Control of the built-in instruments: Of the battery circuits									
4/5	built-in S&H.423820	3 bulbs			0.215 0.215	4.0			3.7	3.0
6/7	built-in S&H.423820	3 bulbs					0.08 0.070	3.4	3.7	3.0
8/9	built-in S&H.423820		0.21 0.2	0.5						

The three built-in instruments show with sufficient precision the series of the three battery circuits

	Control of the instruments in the external circuit									
10 11 12	built-in S&H.423820 S&H.254159	3 bulbs							3.3 3.08 2.95	3.0 2.30
13 14	built-in S&H.423820	2 bulbs								3.15 3.15
15 16	built-in S&H.423820	1 bulb								4.1 4.45

From the built-in tension meter of the external circuit estimations below 3 volt could not be read any more. When slightly over 3 volt the instrument indicates correctly, when over 5 volt, the indication is somewhat too low.

The built-in current indicator of the external circuit points somewhat too high.

The control-instrument S.& H. 423820 is a precision continuous current-instrument (Millivoltmeter) for current and tension measurings. The control instrument H. & B.243159 is a hotwire-instrument.

APPENDIX III

Translation of a Report by Professor W.O.Schumann (Munich) on the examination of the apparatus of Mr. Coler in Berlin on the 19th and 20th March, 1926 (dated 3rd April, 1926)

The apparatus in question principally consists of two parallel connected spools, which being bifilarly wound in a special way, are magnetically linked together. One of these spools is composed of copper sheets (the spool is called the plate spool), the other one of a number of thin parallel connected isolated wires (called: spool winding), running parallel at small intervals to the plates. Both spools can be fed by separate batteries; at least two batteries are necessary to put the spools to work.

The spools are arranged in two halves each, according to the bifilar winding system. To the starting points the batteries are attached, to the parallel connected ends the current-receivers. Besides, inter-communications are connected between parallel windings of the two halves of the plate spool which contain iron rods with silver connections. These rods are magnetised by a special battery through applied windings (called: exciter windings).

According to the statement of the inventor, the production of energy principally takes place in these iron rods, and the winding of the spools plays an essential part in it.

As far as it was possible I convinced myself of the conformity of the circuit with the mechanism. The exciter winding is electrically completely separated from the other windings, which was proved when testing the apparatus while being devoid of tension, as well as when testing it while being at work with the aid of an Ohmmeter and a Millivoltmeter.

In order to ascertain possibly concealed sources, the apparatus was searched with a millivoltmeter, without using any external batteries. No effect was perceptible. Besides the apparatus was carried from the one room through a corridor to another one in the establishment to exclude the possibility of any secret connection to the mains.

Installed in the apparatus were three current meters for the currents from the three batteries, and furthermore current and volt meters of the soft iron type for the current receivers. One and two bulbs respectively were employed for this purpose. Besides there were at disposal one precision milliameter of the S. & H. turn spool type, also employable as ammeter and voltmeter (called: Dr. Sp.C.), and a turn spool millivoltmeter of the A.E.G., an ordinary laboratory instrument, also employable as a voltmeter, (called: Dr. Sp. A.E.G. C.).

While the apparatus fed two bulbs, the current delivery of the three current-supplying batteries was measured directly at the terminal clamps. The results are as follows:

1.	current of the plate battery	48 mA	Dr.Sp. S. & H. C.
2.	current of the spool battery	39 mA	
3.	current of the exciter battery	170 mA	

The indications of the built-in instruments were in unison with the statements of the S. & H. instrument. The total capacity of the batteries of three elements each: 0.257 x 6 - 1.542 watt, (the tension of the batteries in reality being below 6 volt). The possible capacity of the two bulbs according to the built-in soft-iron instruments was 3A x 3.5 v., that is 10.5 watt which means the 6.7 fold of the above-mentioned capacity.

Thereafter only one bulb was connected, and a second one, exactly of the same type, was regulated with a special accumulator battery to get the same light. All currents were measured with the "Dr. Sp. S. & H. C.", all tensions with the "Dr. Sp. A.E.G. C" and not with the built-in instruments.

1. current of the plate battery 28 mA
2. current of the spool circuit battery 23.5 mA
3. current of the exciter battery 180 mA

That means, if reckoned with a 6 volt battery tension, 6 x 0.232 - 1.392 watt. The possible capacity of a bulb, fed by a special battery, and showing the same lightpower was 4.5 volt x 1.5 Am - 6.75 watt. Proportion of both efficiencies ca. 4.85.

Both instruments used for testing (measuring) from S. & H. and from the A.E.G. were then compared with each other by a current-voltmeter, the highest possible deviation being less than 10% which can therefore scarcely influence the result.

The apparatus, according to the statement of the inventor, is adapted now for current increase. The bulb-tension of 3-5 volt is less than the tension of the feeding batteries.

By changing over in the interior it would also be possible to use it for an increase of tension.

Then there was a test with the aid of the Dr. Sp. Millivoltmeter from the A.E.G. to try the decrease of tension in the single winding halves of the plate spool on the right as well as on the left side of the mechanism. (The form of the spool is that of a long small rectangle).

Tensions in Millivolt: Right Side		Winding from the top downward: Left Side	
1. -	2. 0.24	1. 0.34	2. -
3. -	4. 2.6	3. 2.8	4. -
5. 8.4	6. 24	5. 12.4	6. 22
7. 22	8. 24.6	-7. 28	8. 100-140 fluctuates much
9. 26	10. 25	9. 46	10. 30

The tensions are distributed very unequally which must be due to current increases and decreases through the iron-cores.

Further measurements on the single parts could not be carried through for the reason of getting impeded by the built-in parts in the apparatus. Any variations of the regulation-resistances were also not undertaken, because the inventor stated that the apparatus in its installation was very sensitive, especially with regard to the magnetic conditions of the iron-cores, and that a wrong treatment would cause interferences which would be wearisome and very difficult to be eliminated.

The next day I got a Hotwire-voltmeter "H. & B" (called: H.Dr.TH), and further a precision milliameter "S & H." (called: Dr. Sp. TH) at the Technical High School of Charlottenburg. With these instruments and those of the previous day the following tests were made:

1.	Plate circuit	28 mA	Dr. Sp. T H
	Lamp current	1.52 A	Dr. Sp. S. & H. C.
		1.60 A	H Dr. T H
		1.3-1.4	Built-in soft iron instrument
	Lamp tension	4.05 V	H Dr. T H
	Lamp tension	4 V	Built-in soft iron instrument

2.	Spool circuit	28.5-30 mA	Dr. Sp T H
	Lamp current	1.47 A	Dr. Sp S.& H. C
		1.56 A	HDTH
	Lamp tension	3.8 V	H Dr. T H
3.	Exciter circuit	0.173 A	Dr. Sp S. & H. C
	Lamp current	1.5 A	H Dr. T H
	Lamp tension	3.75-4 V	H Dr. T H
	Spool circuit	30-30.5 mA	Dr. Sp T H

Once again a comparison of capacity was carried through by using an equal lamp to that in the apparatus with an accumulator, and bringing it to the same brightness, as judged by the eye.

Lamp in accumulator circuit:

tension	4 V	Dr. Sp S. & H. C
	3.3 V	Dr. Sp A.E.G. C
current	1.5 A	Dr. Sp S. & H. C

Lamp in apparatus:

tension	3.85-4.0 V	H Dr. T H
current	1.59	H Dr. T H

Current in spool circuit 27-28.5 mA Dr. Sp T H

Thereafter the instruments with the measured current - and tension - figures were compared with each other:

1. Tensionmeter parallel on two accumulators:

Dr. Sp A.E.G. C	3.2 V
H Dr. T H	3.8 V
Dr. Sp S. & H. C	4.2-4.3 V

2. Voltaelectrometer in line:

H Dr. T H	1.47 A
Dr. Sp S. & H. C	1.46-1.47

Even when taking into account the errors of the instruments, the resulting multiplication of energy, in principle, does not undergo any alteration through the apparatus. Judging from the conformity of the hotwire - and turnspool instruments in the lamp circuit, the inference is to be drawn that principally it is continuous current one has to deal with. A test of current from the batteries by way of hotwire measurement was not possible.

As a striking fact it should be mentioned that the spool circuit having been at first always switched on alone, received a current of 104 mA. As soon as plates and exciter circuit additionally and simultaneously were switched on, as, according to the inventor, the apparatus demands it, the current in the spool circuit comes down to about 27 mA.

A definite judgment about the apparatus must be reserved by me until all parts have been singly tested, and until variations in the connections in the load and c have been undertaken.

After the present examination, carried through as carefully as the limited possibilities of experimentation permitted, I must surmise that we have to face the exploitation of a new source energy whose further developments can be of an immense importance. The apparatus was visible and accessible in all its essential parts. The inventor agreed quite willingly to each trial in so far as, according to his statement, no harm could be done to the working of the apparatus.

I do not believe in a deception. I deem it expedient to put the apparatus to a further test, and I believe that a further development of the apparatus and an assistance, given to the inventor, will prove justified and of great importance.

APPENDIX IV

Report of Examination on Coler Apparatus

Time: 1.4.43 to 25.9.43

Place: Research Department of the Admiralty (OKM), Berlin, Wannsee

Time: 1.4.43 to 30.6.43

Place: Physical Institute of the Technical University of Berlin

Time: 1.7.43 to 25.9.43

Report by Hans Coler and Dr. Heinz Frohlich

Abstract

Some years ago an apparatus (the Coler apparatus) was empirically developed, which according to a series of reports by well known scientists and practical engineers, was able to deliver a considerably higher electrical output than the input necessary to excite the electromagnetic fields of the apparatus.

Due to the lack of sufficient knowledge of the complicated activity within the apparatus, and of the impossibility of explaining this in known terms, the success in starting the apparatus depended upon happy accident.

With the support of the OKM an attempt was made to examine and measure the activities in the Coler apparatus. Due to lack of sufficient technical and physical means, we did not at first succeed in starting it. This work, however, led to valuable knowledge and information.

It is therefore necessary, through a systematic basic research, to transfer the adjustment and other necessary procedure for starting the apparatus, from the domain of accident into a practical experimental procedure which is at any time reproducible.

1. History

During the years 1923 to 1926 the undersigned, Captain Hans Coler, basing his work on lay theories of electromagnetic and inductive activities, has developed an apparatus for generating electrical energy, which delivered a considerably higher electrical output than was necessary to excite the primary fields of the apparatus. This development was carried out merely empirically and due to the lay procedure, without recording measurements or making diagrams.

This apparatus was the subject of numerous examinations by scientists and practical engineers, who in their reports, almost without exception, had to admit an energy surplus or energy multiplication without being able to explain the process by present knowledge. The inventor also could not give an explanation due to lack of sufficient knowledge of the activities in the apparatus. (See reports by Professors Kloss and Schumann on record at the OKM).

Due to unpleasant differences with the financiers, mostly foreigners, and a nervous breakdown of Coler because of these, the apparatus and original theories were lost.

Upon resumption of the experiments at a later date the effect was occasionally successfully reproduced, (see report by Dr. Modersohn concerning events at New Year 1936-7) which, however, evidentally due to disturbance of the adjustments, could not be maintained because sufficient measurements were not taken. The outbreak of war then put a temporary end to the experimental work.

2. Intervention of the OKM

In 1942 Coler and Dr. Modersohn made an application to the OKM to supply them with materials for further work, hoping that by simple technical means, the apparatus could be made to work.

Their aim was to produce a working apparatus, which could be shown to the Fuhrer of Germany, leaving the measurement of all activites of the apparatus to a later date.

The OKM approved the application and sent sufficient materi to build the apparatus and lent the inventor ordinary amperemeters, voltmeters, and necessary tools. It was found, however that due to the overload of work upon them at this time, their spare time work was not sufficient to secure quick results. The OKM therefore procured their release from other work, to allow a more intensive, and, due to the co-operation of Dr. Frohlich, who had experience in physical measurements, a more scientific approach.

The newly developed apparatus could not be made to work, however, because of the lack of thorough knowledge of the activities in the apparatus which now appeared to be very complicated. Valuable knowledge and promising signs, however, were obtained.

3. Short description of the apparatus

The apparatus consists of three principle circuits which are inter-wound and intercoupled (verkoppelt) in a peculiar way. Some of these are divided again into single subsidiary circuits which evidently have all to be brought into resonance with one another.

The principle circuits (called the anchor) in which the energy gain probably occurs, consists of metal plates between which transformer coils are connected, the whole being connected to one large plate coil (Plattenspule). On each of the single windings on this is coupled a large flat coil (Flachspule) (called field). These flat coils are interwound in two groups, these groups represent the turns of a transformer. This couples, on one hand on the flat coils of the other group (as secondary coils), and on the other hand on the anchor-plate coil, which is placed between them. The third electrically independent circuit (called the directing circuit) (Steuerkreis) regulates this transference. Figure 1 shows diagramatically and in plan these interwound parts (anchor in red, field in green, directing circuit in blue). Figure 2 the so called basic diagram, shows the connections between these different parts. This basic diagram shows the conditions necessary for self-interruption. In consequence of this arrangement, different types of current are created in different conductors (pulsed direct current, alternating current, etc.)

The transformer coils, connected between the anchor plates, are connected in a peculiar way through thin permanent magnet rods. Their main object seems to be to pre-magnetize the transformer cores, it is, however, very probable that they are also connected with the Barckhausen effect, which will be mentioned later.

4. Experiments carried out

From the reflections which led to the diagram of the apparatus described above, the reality of some physical effects was assumed, which partly from the inventor himself, and partly also from the scientists who had examined the working apparatus, were stated as a possible or probable explanation of the phenomenon observed.

As these hypotheses had not been examined by measurements, it seemed suitable, for further development of the apparatus, to clarify these questions by further experiment, so far as was possible with the means at our disposal.

The experiments carried out are described below.

Experiment 1

If a single layer solenoid is connected to a battery, a current begins to flow at the moment of connection, which is known to rise exponentially, until it reaches a constant maximum determined by the voltage of the battery and resistance of the circuit. Every point on the current/time curve corresponds to a magnetic field, which is dependent upon the dimensions of the solenoid. At the moment of connection the electrons do not begin to flow at the same time in all parts of the coil, because the electromagnetic wave (Storungwelle) travels with the velocity of light. The electromagnetic field at different parts of the coil also builds up with a corresponding time difference. Now in a right hand wound coil a north pole is produced at the end where the current enters (the current direction is taken now and hereafter to be the direction of flow of the electrons, i.e. from negative to positive), in a left hand coil a south pole is similarly produced. In the first case the building up of the magnetic field proceeds from the north pole to the south, in the second case from the south to the north, or in other words: The direction of the building up of the field is in one case with the direction of flow of the field and in the other against it.

Question: Is the velocity of the building up of the field different in these two cases, although the data of the coils is the same except for the direction of winding? This means, is there a difference in the time in which the currents are reaching their maximum value?

If such an effect could be found, it must show itself in one of the following ways:

(a) The coils must have different apparent resistances (scheinwiderstand). By measurements of voltage and current the apparent resistance of ten right-handed and ten left-handed coils tested with a Phillips-Schwebungs-Summer. The apparent resistances were the same within an experimental error of about ± 1%.

(b) Other conditions being the same the coils must induce different potentials across a secondary coil, when connected to a battery. Measurements were made with a cathode ray oscillograph from Siemens and Halske. The induced potentials were the same. It can therefore be taken as proved that no such effect exists.

Experiment 2

Given a solenoid consisting of two windings, one upon the other, of the same length and number of turns, enclosing a soft iron cylindrical core. Firmly attached to one end of the core is a pre-magnetized steel rod. If an alternating current is passed through one of these coils, acting as primary, the residual magnetism of the steel rod is strengthened during one half cycle, through the magnetism induced in the core, during the other half cycle it is weakened. If now the other winding is connected in series with the pre-magnetized steel rod - as secondary coil - in such a way that the secondary current must pass through the magnet, one half cycle of the secondary current must be more or less subdued, in other words a rectifying effect must be created. (It may be remembered that according to measurement by Professors Kloss and Schumann a high frequency pulsed direct current - about 180 Kc - is flowing in the output resistance of the apparatus, for the creation of which no other explanation seems possible).

Unfortunately the testing of these effects could only be made with tuned frequency (Tonfrequent) alternating current, with the use of which no such phenomenon could be seen when using measuring instruments and a cathode ray oscillograph.

The question is open whether, and if so in what way, an influence on the electron movement exists through the oscillation of the magnet molecules by high frequency alternating fields, especially of such an impulse-like character.

Experiment 3

Previous measurements, from the year 1937, made by the inventor and his assistant, Dipl. Ing. Rudolf Hingst, had shown the following effect. Given two solenoids each consisting of two coils one upon the other, of the same length and number of turns, one winding of one solenoid is connected in series with one winding of the other, wound in the same sense, and an intermittent direct current is passed through them. The remaining windings, which, are to be considered as secondary coils, are also connected in series but wound in opposite sense to one another. The induced secondary currents are therefore similarly in opposition, and would, due to the equal dimensions of the coils eliminate one another. The measurements referred to above are said to have given the result that in such a secondary circuit a considerable direct current component exists which can be strengthened by means which we shall not go into here. And this strengthening to a degree such that the "secondary direct current" is comparable to the primary current.

Testing the results of these measurements, however, did not confirm them.

<u>Experiment 4</u>

Besides the above questions, uncertainty existed as to what part the above described plate and flat coil unit plays in the function of the apparatus, how great is the mutual influence of the flat coils upon one another? Of the flat coils upon the plates? And finally of the plates between them? As the plates are not only charged as condensers, but also have directed currents passing through them it had to be assumed that their mutual influence not only consisted of a condenser effect, but that they also created a magnetic field.

It must first be stated that the frequency of the "Summer" at our disposal (0-12Kc) was not sufficient to measure the mutual electromagnetic influence of the plates, due to the low apparent resistance. This examination should be carried out with high frequency alternating current. It was, however, possible by the use of a highly sensitive oscillograph as a valve voltmeter to measure the potential induced by the flat coils on the individual plates, to discover the induction currents in the plates, and also to determine the value of the potential induced by the flat coils on the plates in proportion to the currents created in the plates from the current induced in the anchor coils (the directing current was used as a primary winding in these measurements, being connected to the Schwebungs-Summer).

It appeared that the "Ankertrakte" AC or BD (See Fig.1) are not all to be considered as oscillating circuits, but that the single group systems, consisting of plate - to anchor coils - plate, already represent independent oscillation circuits. <u>According to this the apparatus contained ten such oscillation circuits.</u> The individual frequency of these circuits <u>and possible differences existing between them</u> could unfortunately not be measured, because, as already mentioned, up to now only tuned frequency for excitation was at our disposal. In order to make the apparatus work the harmony of all oscillating circuits in their individual frequencies would evidently be necessary - at least within certain limits determined by the suppression in question. Such an adjustment can of course not be secured by the means at our disposal at the present. The previous occasional success must be considered as due to chance.

The mutual influence of the flat coils upon one another could on the other hand easily be examined because, despite the great distance between the windings (25 mm), they have remarkably great apparent resistance (about 200 ohms at 10 Kc).

It appeared that the power factor (Ubertragungs-faktor) of both of the flat coils wound 1 : 1, in consequence of their peculiar interwinding (See Fig. 1) have the astonishingly high value of 0.85. The value of the power factor was at a maximum at 10 Kc, at which frequency the most favourable matching of the impedence of the Summer was obtained.

The assumption could easily be made that the power factor of the flat coils upon the plates is still more favourable, but this measurement can only be made with high frequency alternating current.

Apart from the great number of alternative arrangements and connections between the different parts, - during the period covered by this report eight different circuit diagrams were tested in addition to the experiments above - the following so far unsolved problems are hindering success in making the apparatus work:

(a) The influence of the pre-magnetization and of the magnetizing effect due to the battery current, upon the individual frequency of the oscillating circuits.

(b) The influence of the Barckhausen effect on the phase conditions of the oscillations in the individual circuits and on the mutual magnetic and electrical processes.

(c) The influence on the movement of electrons in the pre-magnetized steel rods by the high frequency pulse-like field variations, through the oscillations of the molecules of the magnets.

(d) The behaviour of the mutual effect between flat coils and plates in a high frequency system.

5. Conclusions

After experience previously gained on the working apparatus, and on the basis of the technical measurements and examinations mentioned above, the picture of the way in which the apparatus works is as follows.

Due to the connecting of the batteries, a current impulse is induced in the anchor circuit which charges the plates. The discharge current from the plates causes electrical interruption of short duration of the battery current in the field circuit, which furthermore inductively interrupts, or changes the direction of, the battery current in the directing circuit for a short time. The electromagnetic field induced by this process in the directing circuit by its dissipation, induces over the field circuit a current in the anchor circuit recharging the plates, and so forth.

Due to the influence of the Barckhausen effect each single process has an impulse-like character, and the necessary change of phase is produced to allow the regularity of the process.

Due to a source, up to the present not investigated, and not explainable by existing scientific theories, an additional quantity of energy is freed during each cycle which leads to a continual raising of the amplitude of the mutual processes, until the magnetic cores are saturated.

From the fact that in the resistance of the apparatus pulsed direct current is flowing (see Report by Professors Kloss and Schumann) there is possibly an up to now unknown rectifying effect, or alternatively the gain in energy is produced only during one half of the cycle, either during the charging or discharging of the plates. The activity in the apparatus must take place <u>in the ten oscillation circuits in a phase-like manner</u>. As mentioned above no technical means were available to make the necessary tuning adjustments.

It is clear from the above that the success of the inventor up to now could only be due to chance, or happy accident. The necessity, therefore, arises to transfer the apparatus from the state of empirical development, with sufficient technical means and based on results of an exact basic research, to a state of working procedure which can be controlled.

6. Further Procedure

In a simple apparatus, possibly only consisting of one oscillation circuit, the problems mentioned in paragraph 4 should be cleared up. The excitation of the oscillation circuit being caused by an appropriate valve circuit. After clearing these problems a second oscillation circuit may be added, which will allow the examination of the conditions necessary for the mutual building up of effects.

It will then be possible for the first time to go from external excitation to self interruption and a solely electromagnetic basis, with the aim of bringing the apparatus in question to work systematically.

Berlin, 27 September 1943

(Coler) (Dr. Frohlich)

APPENDIX V

Extract from Statement by Coler relating to Frohlich's experiments

These fundamental researches, which have made the first real and large breach in the citadel of present scientific belief, will now be described in detail. The installation shown in Figure1-4 was used in the experiments. A transformer, with a ratio of 1 : 1 was connected on its primary side through a periodic make-and-break X to a D.C. source U. Great care was taken, to ensure that the make and break functioned smoothly, without backlash and uniformly, at high frequencies and thus produced as clear and unambiguous a result as possible.

On the secondary side the coil S was connected with resistance R, a reversible rectifier G, and a highly sensitive hot-wire ammeter, to form a circuit. The resistance R was large compared with the apparent resistance of the coil S.

Switch Sch allowed the switching of the rectifier G to be carried out in such a manner, that in one case the half-wave of the opening impulse, and in the other that of the closing impulse, was transmitted. In order to prevent any possible objection, that the iron had any influence on the results, the whole transformer was cut out during the experiments and an iron-free flat coil arrangement used as the inductor during these experiments. This eliminated all interference and had excellent transmission properties. The required sensitivity was attained in these measurements by using a thermopile with the most sensitive milliammeter. After many trials, the motor-driven interrupter of a Bosch ignition magnet was used as make-and-break. This gave very uniform and smooth interruption with a frequency of 100 c/s. The result obtained with this experimental arrangement, which already eliminated all the possible causes of later objections, was the clear proof of a considerably larger energy during opening (intake), compared with closing impulses. The energy was determined by i^2 from the measured current intensity, and as the resistance R is large compared to the apparent resistance of the coil, the objection, that the opening impulse had a frequency spectrum of considerably higher value than the closing impulse, could not hold good.

In order to counter all other plausible and possible objections the processes were recorded by an oscillographic method. Unfortunately the light strength of the available cathode ray oscillograph tubes (Braun's tubes - it was during war time) did not suffice to produce perfect oscillogramms. Consequently, for the time being at least, this aid, which would have completely removed all objections, had to be dispensed with and a Siemens' loop oscillograph was therefore used. By choosing suitable loops, sources of error were eliminated as far as possible, or at least kept within very small limits.

The most careful planimetric measurements again gave the same results as those obtained with the experimental installation described above, at least in principle, even though the percentage difference of the quantity of energy in the opening and closing impulses was slightly less in the measurements of the oscillogram than those obtained from the thermopile. It still amounted to an average of 10%, compared with 50% in the measurements, so that errors in the planimetric measurements are not to be assumed. Also, if this had been merely a scattering of the results, a difference in favour of the closing impulse must have occurred at least once. However the difference was always in favour of the opening impulse. Any possible errors in the planimetric measurements are controlled and avoided, by carefully cutting out the figures produced and weighing them on a most sensitive and accurate balance.

Consequently this observation can also be considered as a proof of the fact that an energy difference exists. The circumstance that the percentage of this unequality appears smaller in the oscillographic measurements than in the thermal, which at a power frequency of 100 c/s assuredly yield a sufficiently accurate mean value, allowing also for the resistance ratios of the rectifier, can readily be explained by the lag of the oscillograph used, which is not even capable of recording the considerable, but extremely short, energy peak of the opening impulse. Although, therefore, the inertialess cathode ray tube was not used in these measurements as a final proof, I can consider my discovery of the energy difference between the opening and closing impulse as proved on the basis of Frohlich's experiments.

As this also proves simultaneously, that my intuitively derived view, based on my most primitive experiments, of these processes has proved correct, the reader will forgive me, if I consider my other two results as correct, until new, and above all better explanations are found for the phenomena described. This all the more, as my development of the "Space energy receiver" was based on this and was successful.

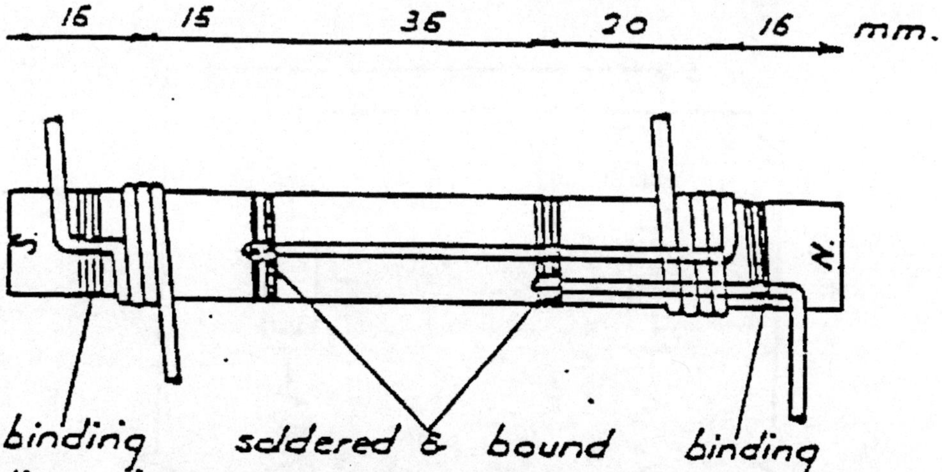

Note: Paper insulation between magnet and coil.

Fig: 1.

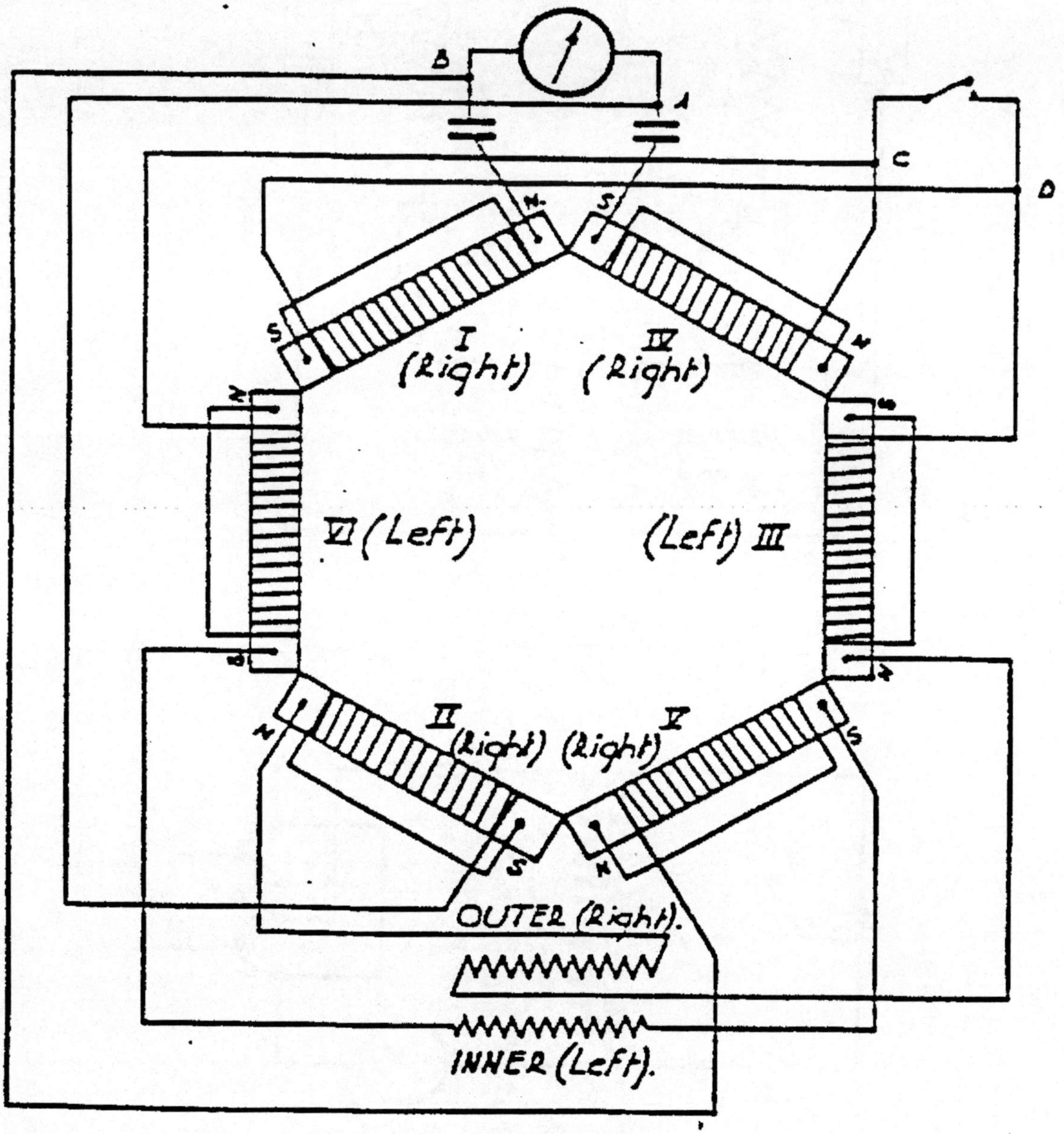

Fig: 2.

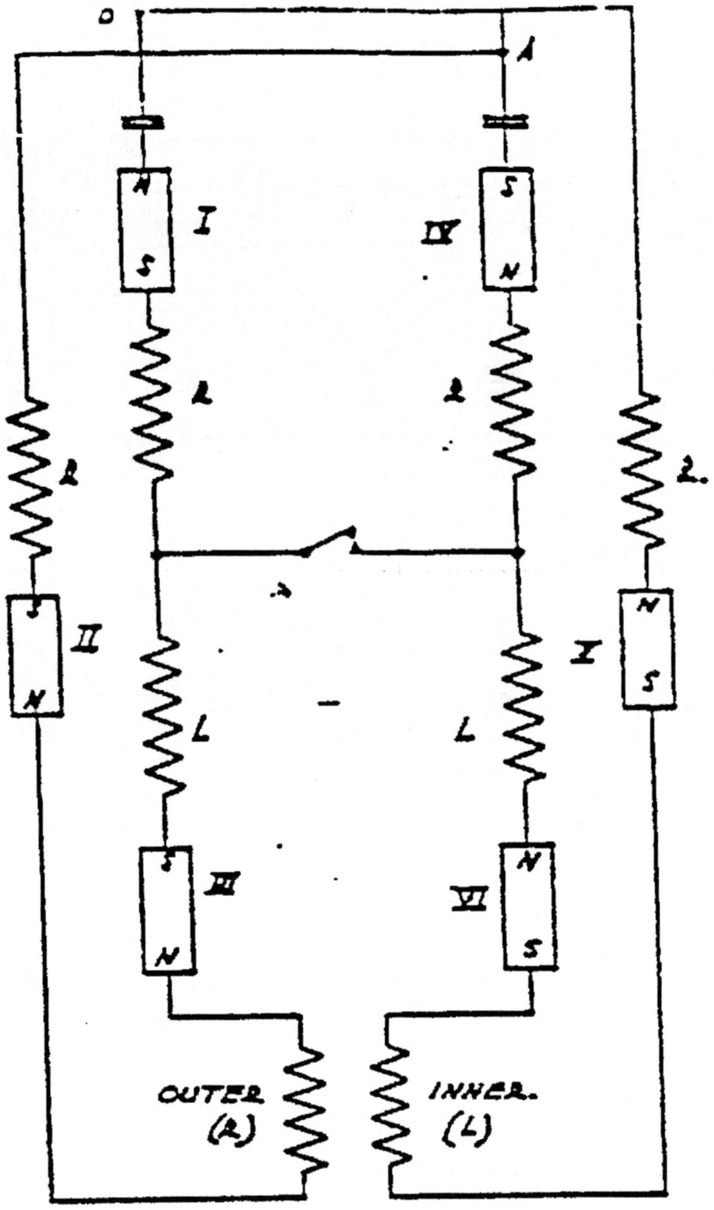
O
A
N
S
I
S
N
IV
R
R
R
R
S
N
II
N
S
V
L
L
S
N
III
N
S
VI
OUTER
(R)
INNER.
(L)

Fig. 3.

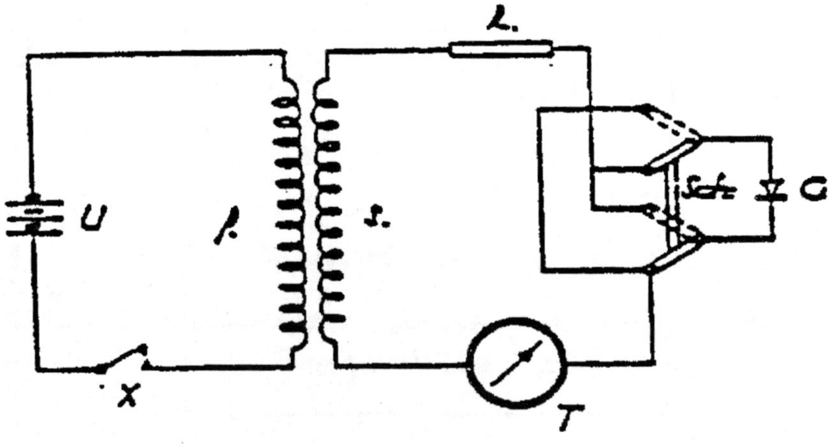
R.
U
P.
S.
Sch
G
X
T

Fig. 4.

Experiments with Coler Magnetic Current Apparatus

Thorsten Ludwig

German Space Energy Association (DVR) and Berlin Institute for Innovative Energy and Propulsion Technologies (Binnotec), Bouchéstr.12
12435 Berlin, Germany
+491716280357; DrLudwig@thorstenludwig.de

Abstract. Experiments with a replica of the famous Coler "Magnetstromapparat" (magnetic current apparatus) were conducted. The replica was built at the same institute at the Technical University of Berlin where the original was tested by Prof. Kloss in 1925. The details of the setup will be presented in this paper. The investigation of the Coler device was done with modern methods. The output was measured with a digital multi meter (DMM) and a digital storage oscilloscope (DSO). The results of the measurements will be presented. Did Coler convert vacuum fluctuations via magnetic, electric and acoustic resonance into electricity? There is a strong connection between magnetism and quantum field radiation energy. The magnetic moment of the electron is in part an energy exchange with the radiation field. The energy output of the Coler apparatus is measured. Furthermore the dynamics of the ferromagnetic magnets that Coler reported as the working principle of his device was investigated with magnetic force microscopy (MFM) and the spectroscopy mode of an atomic force microscope (AFM). The magnetic and acoustic resonance was investigated with magnetic force microscopy (MFM). The connection between ZPE and magnetism will be discussed as well as the perspective of using magnetic systems as a means to convert vacuum fluctuations into usable electricity.

Keywords: Magnetism, Coler, Magnetic Force Microscopy, MFM, Vacuum Fluctuations, Zero Point Energy, ZPE, Atomic Force Microscopy (AFM)
PACS: 41.20.Gz, 75.20.En, 75.25.-j, 75.30.-m, 75.30.Ds, 75.30.Fv, 75.30.Kz, 75.80.+q, 88.90.+t

INTRODUCTION

This article will explore a new type of electrical generator which was invented by Hans Coler. First the history and the general description of the device will be given. The historical part will also mention the early tests of the device by the well known German Professors Schumann and Kloss, which are the main reason why this invention is not forgotten. Another important historical fact is the investigation of the device by the British Intelligence Objectives Sub-Committee (B.I.O.S.) and the confidential, later disclosed, positive report of this agency. The investigation was conducted by Hurst and Sandberg. Hans Coler was contracted in 1947 to rebuild his magnetic current apparatus. This work will present the technical design of a replica of such a Coler device and measurements on that replica. The replica was built by students of the Technical University of Berlin in 1998 and 1999 under the supervision of Dipl. Ing. Andreas Manthey. The construction of the Coler device was part of a course in interdisciplinary innovation learning. The student group took up the Coler device because there is a rich source of technical information on the device and another important reason was that the original device was positively tested at the same institute in 1926. An investigation to retrieve the original documents in the archives of the Technical University revealed that the documents were a toll taken by a British firebomb during a air raid on Berlin in 1943. Since Coler also presented some theoretical hints on the working principle of his device, this work will also deal with the fundamental properties of magnetism and the feasibility of using electrical and magnetic resonances of ferromagnetic magnets to construct a generator. This article will also present an experimental technique that allows testing Coler's statements on magnetism. The work on magnetism is still ongoing. This work on the Coler generator was taken up because of the credible witnesses for a proof of principle for a device that works with permanent magnets and no other input power. Since the magnetism of the electron can be explained as a relativistic quantum effect with corrections due to the exchange of energy with the zero-point energy quantum fluctuations and permanent magnetism is today seen as a collective electron phenomenon, Coler seems to be good starting point for practically testing the possibility of using magnetism to convert ZPE into useable electric energy. This connection will be laid out in more detail in a later part of this paper.

A major part of the work presented here is the investigation into the nature of magnetism with modern methods such as magnetic force microscopy (MFM) and the origin of magnetism (Berman *et al.*, 2006). Permanent magnets are today understood as a collective phenomenon of the electron spin and the magnetic moments of the electrons involved (Auerbach, 1994). The magnetic moment of the electron is a quantum field phenomenon (Maggiore, 2005).

In order to unveil the working mechanism of the Coler generator two approaches are combined in this paper. On the one hand it was tried to replicate Coler´s magnetic current apparatus and to conduct modern measurements on the replica. In addition another avenue was followed to single out the working principle with modern methods to find the resonance between ferromagnetism and electrical oscillations that Coler gives as a reason for output energy of his device.

THE COLER GENERATOR

Hans Coler was the inventor of two devices by which electrical energy may be derived without a chemical or mechanical source of power. The devices are called the "Magnetstromapparat" (magnetic current apparatus) and the "Stromerzeuger". The "Magnetstromapparat" was developed by Coler and von Unruh early in 1933. With this device, consisting only of permanent magnets, copper coils, and condensers in a static arrangement he showed that he could obtain a tension of 450 mV for a period of some hours. This device consists of six permanent magnets wound in a special way so that the circuit includes the magnet itself as well as the winding. These six magnet-coil combinations are arranged in a hexagon and connected in a circuit which includes two small condensers, a switch, and a pair of solenoid coils, one sliding inside the other (Hurst, 1956).

The "Stromerzeuger" consists of an arrangement of magnets, flat coils, and copper plates, with a primary circuit energized by a small battery. Coler and von Unruh made up a slightly larger model with an output of 70 watts. This was demonstrated to Dr. F. Modersohn, who obtained from Prof. Schumann and Prof. Kloss confirmation of their tests in 1926. The device was seen by the professors Schumann (Munich), Bragstad (Trondheim) and Knudsen (Copenhagen). Reports by Prof. Kloss (Technical University Berlin) and Prof. Schumann (Technical University Munich) are credential evidence that the device was real and worked. Prof. Schumann is still widely known for the discovery of the Schumann Resonances in the earth atmosphere. Coler then in 1937 built a lager version with an output of six kilowatts.

Since an official interest was taken in his inventions by the German Admiralty the British intelligence started an investigation, although normally it would be considered that such a claim could only be fraudulent. Coler was visited and interrogated. He proved to be co-operative and willing to disclose all details of his devices. He built a "Magnetstromapparat" in 1947 using material supplied to him by the British intelligence (Ministry of Supply), and working only in their presence. The greatest tension obtained with this "Magnetstromapparat" was stated to be 12 volts.

Coler had no complete scientific explanation for the working of his device. He stated that his researches into the nature of magnetism had lead him to conclude that ferromagnetism was an oscillating phenomenon, of a frequency of about 180 kilohertz. This oscillation took place in the magnetic circuit of the apparatus, and induced in the electrical circuit oscillations the frequency which of course depended on the values of the components used. These two phenomena interacted, and gradually built up the tension. Coler stated that the strength of the magnets did not decrease during use of the apparatus; and suggested that he was tapping into a new sort of energy hitherto unknown -"Raumenergie" (Space-energy).

Technical details of the Coler replica used in this work

The Coler generator consists of six permanent magnets with coils wound around it and connected to it in a special way. The electric circuit therefore includes the magnets as well as the windings. The six magnet-coil combinations are arranged in a hexagon and connected as shown in Figure 1.

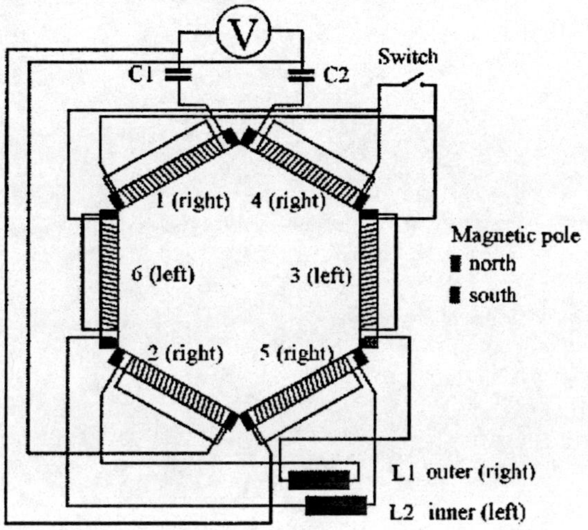

Figure 1. Electric diagram of the Coler magnetic current apparatus.

The circuit includes two small capacitors, a switch, and a pair of air coupled solenoids (L1 and L2) one sliding inside the other. The values of the capacitors were 4.7 nF each. The Coler replica used in this work is shown in Figure 2.

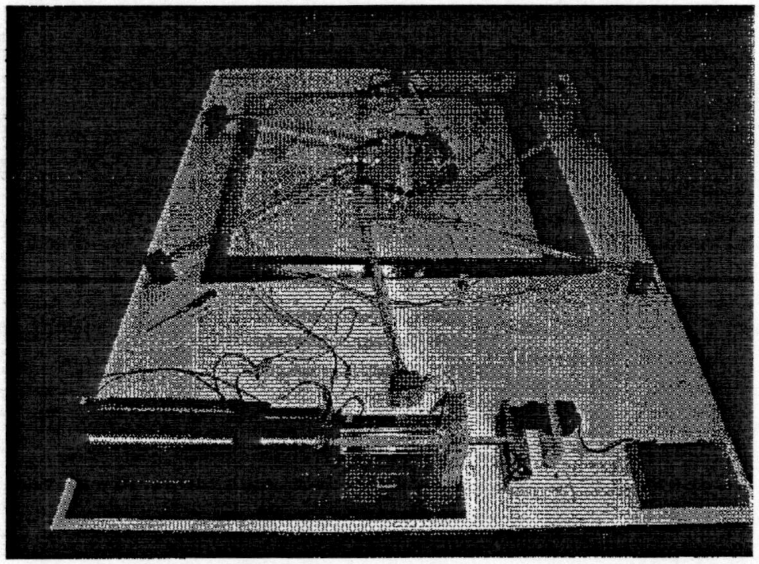

Figure 2. Coler replica used in this work.

One of the important uncertainties of the Coler generator are the material of the magnets and the construction and binding of the coil on the magnets. Some features seam to be schematic other not. It is unclear if the coils should be wound over most part of the magnets as in Figure 1 or only on the edges. The student team decided to wind a homogeneous coil that is soldered and bind as described in the British intelligence report, as seen in Figure 3. The

Magnets were especially manufactured for the original TUB project by IBS Magnet in Berlin. The length of the magnets is 100 mm and the average remanent magnetisation is 200 mT. The magnets were cut and turned to a length of 100mm and a diameter of 15 mm out of a special pure iron. After machining the iron rods were magnetized by IBS Magnets.

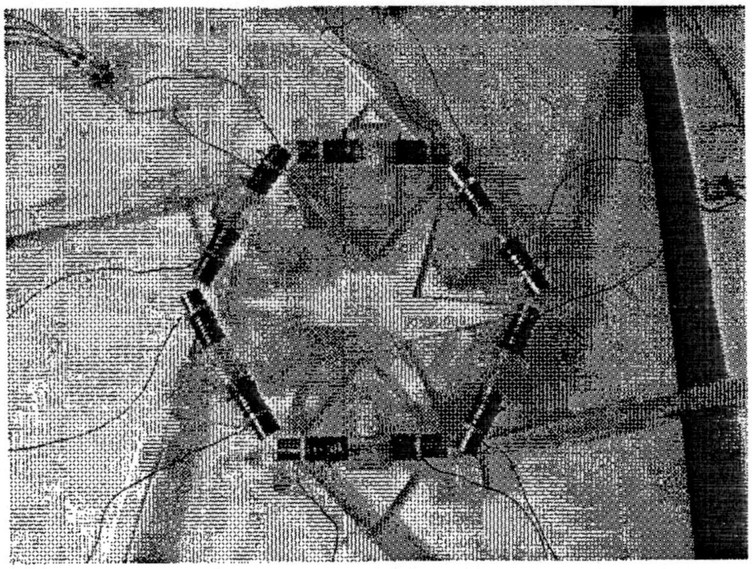

Figure 3. Close up of the magnet hexagon.

The stepper motors seen in Figure 2. are part of the original concept design of the Coler replica. It was planned in 1998 at the TUB to let a computer program find the optimal position of the magnets automatically. The computer and the program could not be retrieved. Electrical and magnetic properties of the permanent magnets and coils used in these experiments are shown in Table 1. Numbers correspond to numbers of the magnets-coil combinations in Figure 1.

Table 1 Electrical and magnetic properties of the permanent magnet-coil combinations.

Magnet-coil	Resistance	Magnetic field at pole
1	0.386 (0,001) Ω	N: 200, S: 180 mT
2	0.232 (0,001) Ω	N: 180, S: 180 mT
3	0.440 (0,001) Ω	N: 170, S: 200 mT
4	0.310 (0,001) Ω	N: 200, S: 180 mT
5	0.385 (0,001) Ω	N: 200, S: 200 mT
6	0.416 (0,001) Ω	N: 180, S: 200 mT

INVESTIGATION OF MAGNETIC AND ELECTRIC PROPERTIES IN THE COLER "MAGNETSTROMAPPARAT"

In the report of the British intelligence the following normal start up procedure for the Coler apparatus is given. Before the device is used, it is kept in the zero position for some time. In this position the switch is open and every magnet is touching the next one. To bring the device into operation, the switch remains open, the magnets are moved slightly apart, and the sliding solenoids are set into various positions, with a pause of several minutes between adjustments. The magnets are then separated still further, and the sliding solenoids are moved again. This process is repeated until at a critical separation of the magnets a reaction in form of a higher voltage is seen on the voltmeter. The switch is than closed, and the procedure is continued more slowly. The tension is then suppose to build up to a maximum where it remains.

A maximum voltage of 133 mV was observed. In the closed position the output voltage was 0.012 mV ac and below 0.00 mV dc. In most positions of the magnets the output voltage was 2-3 mV ac and dc. In a few cases the voltage went up for some minutes to 15 mV, 20 mV, and 46 mV ac with 6 mV DC. In one case the voltage remained for about 5 minutes at 133 mV. An overview of exceptional high output voltages is given in Table 2.

Table 2. Output voltages.

Time	Output Voltage AC	Output Voltage DC
Closed loop	0.012 mV	0.00 mV
Usual output	2-3 mV	2-3 mV
2-3 min	15 mV	
3 min	20 mV	5-10 mV
3 min	46 mV	6 mV
5 min	133 mV	

The exceptionally high output voltages did not last long enough to close the switch and to go on with the next run up stage. The magnetic field strength at the time of the highest output voltage is shown in Table 3.

Table 3. Magnetic field strength

Position, between	Magnetic Field Strength
1-6	310 mT
6-2	380 mT
2-5	210 mT
5-3	280 mT
3-4	310 mT
4-1	300 mT

The output voltage was measured with a Fluke 189 true RMS multi meter and recorded with digital store oscilloscope DSO-5200A from Voltkraft. The digitization rate was 250MS/s at a 9 bit resolution. The results of the measurements with the Fluke DMM are shown in Table 2. The recording for the 46 mV case is shown in Figure 4.

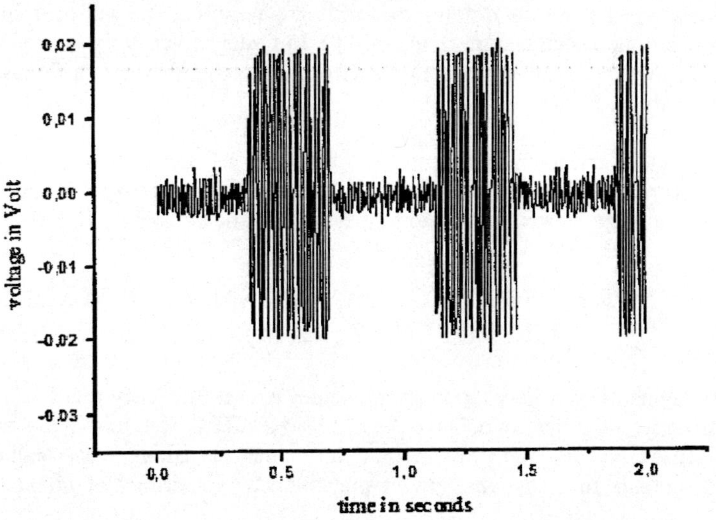

Figure 4. Output voltage over time in the case of 46 mV output voltage.

MAGNETISM AND QED

The magnetic moment μ_s of the electron is given by

$$\mu_s = -g_s \frac{e}{2m_0} s \quad (1)$$

In equation (1) g_s is the gyromagnetic ratio, e is the elementary charge, s the spin quantum number and m_0 the mass of the electron. In the classical case g is expected to be 1.

The experimental value for g given by Dehmelt (1990) is 2.02319304376(8).

It is one of the most important achievements of the relativistic quantum mechanics that the Dirac equation for an electron in an external field produces almost the right magnetic moment of the electron. The expert is not surprised by this as the magnetic field can be seen as a relativistic effect of the electric field (Schmueser, 1988). The Dirac equation results with a g_s of 2. The difference of the experimental value to g_s is called g-2. The difference can be calculated to a high precision with QED. This process is called radiation correction, as the difference g-2 is the result of energy exchange with the vacuum fluctuations, also called radiation field or zero point energy. Schwinger was the first one to show that the magnetic moment can be calculated from QED (Schwinger, 1948). His result for g-2 is α/π, with α being the fine structure constant. Luttinger showed a simpler way to arrive to this result (Luttinger, 1948). Masperi gives a more qualitative and understandable explanation (Masperi, 1989). These calculations have been experimentally verified for example by the single electron experiment of Dehmelt (1990).

The modern view of the magnetic moment of the electron is that it can be calculated from quantum field theory. The magnetic moment of the electron is partly connected with the vacuum fluctuation because the g-2 part is caused by interaction with the radiation field. Permanent or remanent magnetism is today seen as a collective phenomenon of single electron magnetic moments (Auerbach, 1994). It is therefore justified to investigate permanent magnetism as a means to converts vacuum state energy, into other forms of energy, *i.e.*, electricity. Puthoff and Cole have shown that in principle this energy can be extracted from the vacuum (Puthoff and Cole, 1993). Coler might have found a mechanism that converts zero point energy into usable electricity. He points out that he uses a special resonance of magnetic and electrical vibrations in ferromagnetic material, especially iron. Since ferromagnetism is a collective effect, it can be seen as a self organized dissipative structure. This holds especially for the ferromagnetic resonance. Therefore ferromagnetic materials combine fundamentally vacuum fluctuations and dissipative structures. The self structuring process of the electrons in ferromagnetic material could be a way to cohere vacuum fluctuations. This would make ferromagnetic materials ideal candidates to fulfill the requirements laid out by Moray King for processes that extract energy from the vacuum state (King, 1989). In Order to verify the connection between Coler magnetic current apparatus, magnetism and energy extraction from the quantum vacuum field it is planned to use Casimir force measurements (Ludwig, 2008).

In order to understand Coler´s device and to engineer technology from it, a better understanding of the dynamics of ferromagnetism is needed. Research in magnetic nano-structures, ferrofluids, single spin magnetic resonance force microscopy and magnetic semiconductors promise to lead to these needed new insights.

TEST OF COLER´S THEORY OF FERROMAGNETISM WITH MAGNETIC FORCE MICROSCOPY (MFM)

Coler´s suggestion that ferromagnetism is a fluctuation phenomenon with a frequency around 180 kHz was studied using magnetic force microscopy. In order to achieve this; an ANFATEC standard AFM was equipped with chromium-cobalt coated silicon cantilevers. With this method very small variations in the magnetic properties can be detected. The method is suitable to study changes in the magnets. Experimental method and data will be presented in this section.

Magnetic Force Microscopy

Magnetic force microscopy (MFM) is a special mode of atomic force microscopy (AFM). In MFM a ferromagnetic tip is scanned at a certain constant height of several ten nanometres above the surface. This mode is also called fly mode. In fly mode first the line is scanned in AFM dynamic mode and then the same line is scanned at a height of several ten nanometers above the topography just measured in the AFM mode. At that height there is no atomic force between the tip and the surface but the strength and orientation of the magnetic field of the surface can still create a force on the ferromagnetic tip. Figure 5 (a) shows the working principle of an AFM. A thin silicon cantilever is scanned line by line over a sample surface. The nm resolution for the scan is achieved with a piezo actuator. The piezo actuator can be positioned in 3 dimensions (xyz) with medium voltage. The force on the tip and therefore on the cantilever is measured with a laser beam that is reflected on the back of the cantilever and a segmented photodiode. If a force is supplied to the cantilever, the cantilever bends, which lead to a shift of the position of the laser beam on the photodiode. If the image is acquired with the tip touching the surface it is called contact mode AFM. If the cantilever is vibrated with tip just above the surface and the force is measured as a change in amplitude or phase shift between driving frequency and photodiode signal it is called dynamic mode AFM. For MFM the AFM is usually used in dynamic mode. Figure 5 (b) shows the MFM imaging with the fly mode. Figure 5 (a) shows the working principle of an AFM which is used in MFM to record the surface topology and Figure 5 (b) shows the working principle of the MFM mode of an AFM which is used for imaging the magnetic properties of the sample. To scan the magnetic properties the cantilever and tip are coated with a Cr-Co layer.

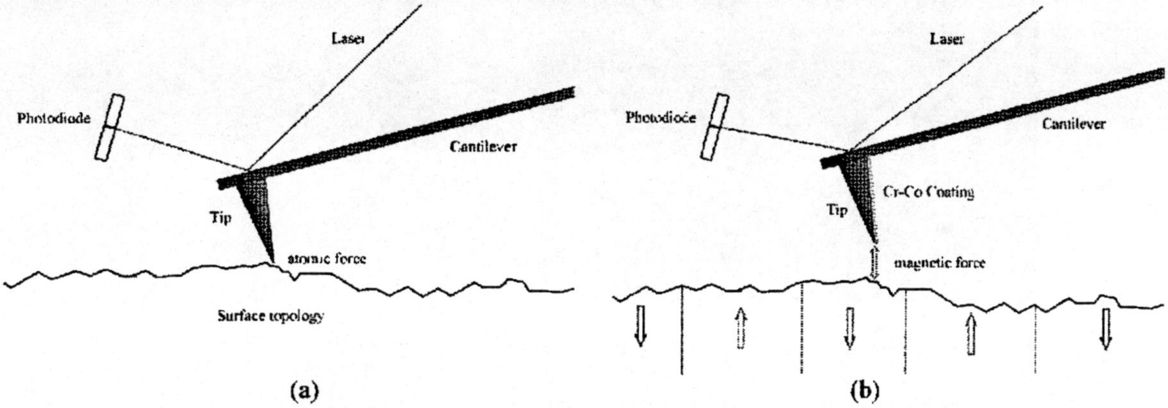

Figure 5. (a) Working principle of an AFM, and (b) of the MFM mode.

The AFM used for the here presented work is shown in Figure 6.

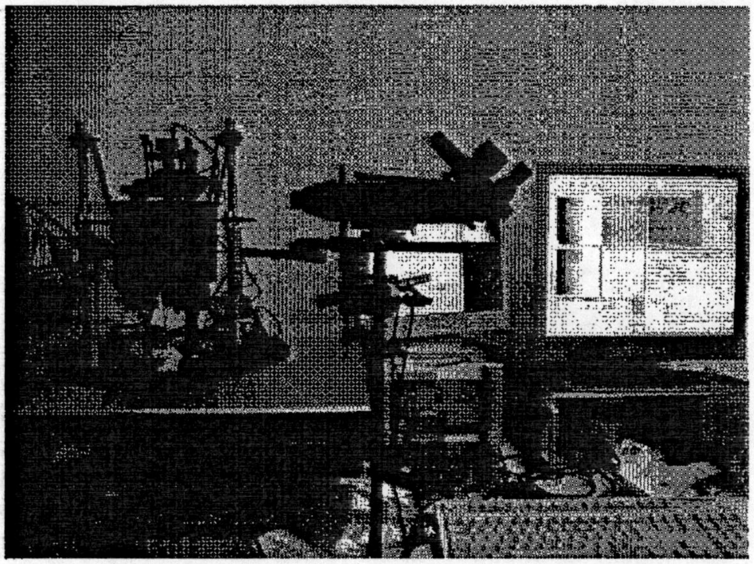

Figure 6. The atomic force microscope used in this work.

In our case the silicon cantilever is 230 µm long, 40 µm wide and 3µm thick. The resonance frequency is 81 kHz and the force constant is 3.5 N/m. The tip and both sides of the cantilever are coated with continuous films of cobalt 1st layer with a thickness of 60 nm and a 2nd layer of chromium with a thickness of 20 nm. The Cr layer is for protecting the cobalt layer chromium. Figure 7 shows a MFM image of a test structure. In the MFM image there is no trace of the topography, which is shown in Figure 8. The topography image was taken with the normal dynamic mode of the AFM. The MFM image shown in Figure 7 was taken in fly mode at a height of 45 nm. The test structure was a piece of a ZIP 100 MB disk. The scanned area was 8 by 8 µm. The tip has a radius of less than 10 nm. The resolution of the images is 5 – 10 nm.

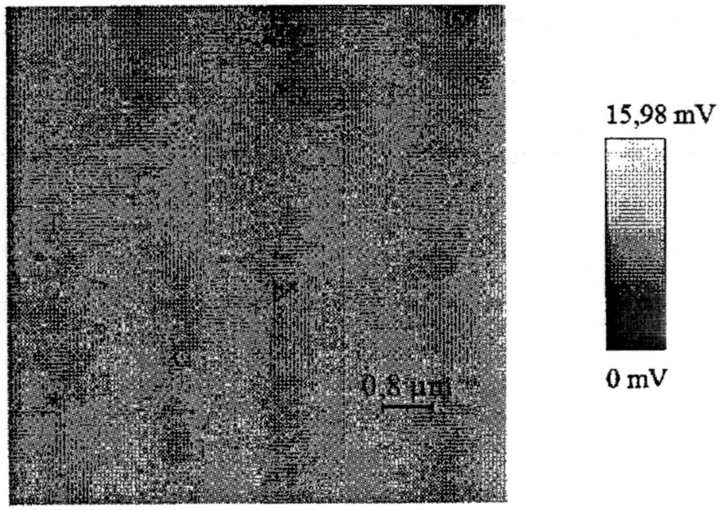

Figure 7. MFM image of the magnetic field strength of a test structure

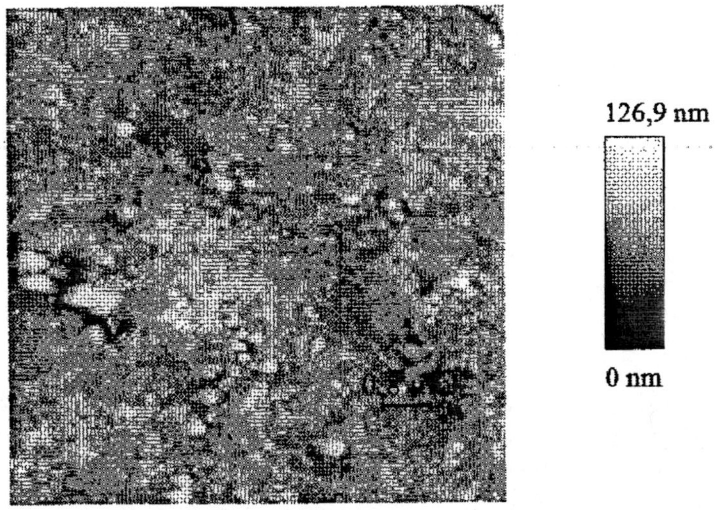

Figure 8. Topography image of the same area as in Figure 7 acquired with AFM in dynamic mode.

MFM Study of Iron Samples

In order to study Coler's theory iron was scanned with MFM to acquire pictures of the magnetic strength distribution and its dynamics. In Figure 9 an image of an iron plate is shown. In this case the phase angle between driving force and the photodiode signal was used for contrast.

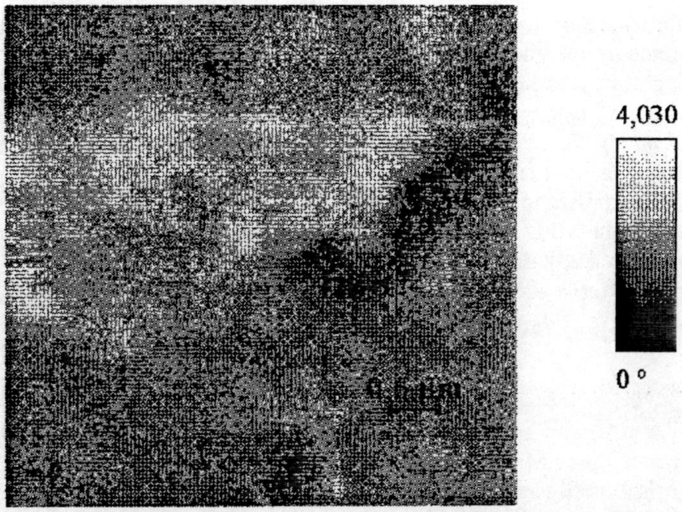

Figure 9. Magnetic field strength of iron in phase contrast.

In order to verify Coler's hypothesis of the 180 kHz oscillation in ferromagnetic material the spectroscopy feature of the AFM was used. In this case the tip with Cr-Co coating was positioned just above the surface and the photodiode signal was scanned from 20 kHz to 400 kHz. The result of this frequency scan is shown in Figure 10.

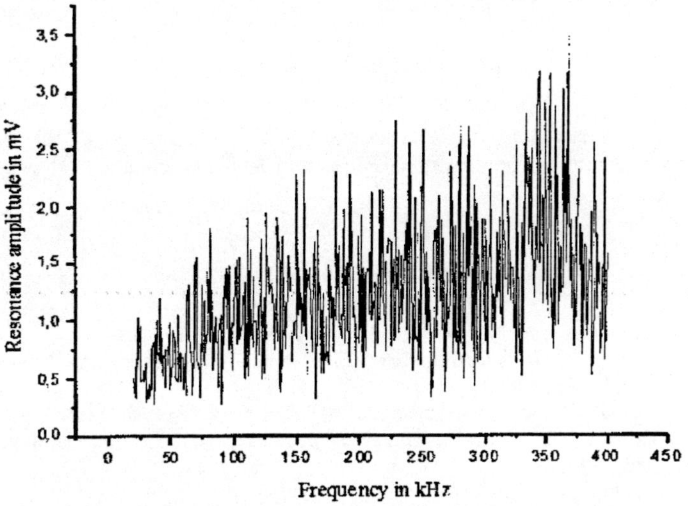

Figure 10. Frequency scan of ferromagnetic iron sample.

So far no clear evidence of a ferromagnetic oscillation phenomenon around 180 kHz could be found. The presented data are to be seen as samples to demonstrate the experimental methods used. It constitutes by no means a complete investigation of ferromagnetism. The investigation of iron samples and iron permanent magnets has to be continued. Of special interest is the measurement of permanent magnets in the magnetic and electrical resonance in Coler's circuit. This still remains to be accomplished.

CONCLUSIONS

The origin of excess power obtained via magnetic resonance was investigated in the case of a replica of the Coler generator. Even though a maximum of 133 mV and at times 20 to 40 mV could be detected the conclusion is that the

Coler magnetic current apparatus could not be set into function as it was reported by Hurst in the British intelligence report. Since Hurst and Coler point out that it is difficult to hit the right resonance position this is not surprising. The work on the replica is not completed. There are still a number of parameters that can be varied. In addition it is planned to do the same measurements on another replica from a Technical University. The AFM and MFM measurements have also not brought a breakthrough insight into Coler's claims. The resonance frequency of 180 kHz was investigated by the means of magnetic force microscopy. Therefore no clear mechanism on how the quantum field fluctuations could supply the energy to drive the circuit via the magnetic moment of the electron could be identified to this day. It has to be mentioned that the work is still ongoing and it is hopped to refine the measurement techniques and sample preparation to find definitive results in the future. It is planned to study the same permanent magnet material with the MFM that is used for the magnets in the circuit. Furthermore it is planned to prepare smooth pure iron samples to study the ferromagnetic dynamics in iron. It seems promising to follow up the methods laid out in this work to learn more about the nature of ferro- and permanent magnetism and the possibility to convert radiation field energy into electricity.

ACRONYMS

AFM – Atomic Force Microscope
B.I.O.S. – British Intelligence Objectives Sub-Committee
DSO – Digital Storage Oscilloscope
DMM – Digital Multi Meter
MFM – Magnetic Force Microscope
QED – Quantum Electro Dynamics
RMS – Root Mean Square
TUB – Technical University of Berlin
ZPE – Zero Point Energy

ACKNOWLEDGMENTS

The authors thank the Deutsche Vereinigung für Raumenergie (DVR) for support of this work. The author also thanks Tom Valone and Marco Bischof for helpful discussions and Andreas Manthey for providing the replica and documents on the earlier project.

REFERENCES

Auerbach, A., *Interacting Electrons and Quantum Magnetism*, Springer, New York, (1994).

Berman, P. G., Borgonovi , F., Gorshkov V. N. and Tsifrinovich, V. I., *Magnetic Resonance Force Microscopy and a Single-Spin Measurement*, World Scientific Pub., Singapore, (2006).

Dehmelt, H., "Experiments with an Isolated Subatomic Particle at Rest," *Rev. Mod. Phys.*, **62**, (1990), pp. 525-530.

Hurst, R., "The Invention of Hans Coler, relating to an alleged new Source of Power" *B.I.O.S.*, Final Report No.**1043**, Target Number: C31/4799, [declassified], (1956).

King M., *Tapping Zero Point Energy*, Paraclete Publishing, Provo UT, (1989).

Ludwig, T., "Casimir force experiments with quartz tuning forks and an atomic force microscope (AFM)" *J. Phys. A Math. Theor.*, **41**, (2008), p. 164025.

Luttinger, J. M., "A Note on the Magnetic Moment of the Electron," *Phys. Rev.*, **74**, (1948), p. 893.

Maggiore, M., *A Modern Introduction to Quantum Field Theory*, Oxford University Press, Oxford, (2005).

Masperi, L., "Qualitative description of the electron's anomalous magnetic moment," *Revista Brasilleira de Fisica*, **19**, (1989), p. 215.

Puthoff, H. E. and Cole D. C., "Extracting energy and heat from the vacuum," *Phys. Rev. E*, **48**, (1993), p. 1562.

Schwinger, J. S., "On Quantum-Electrodynamics and the Magnetic Moment of the Electron," *Phys. Rev.*, **73**, (1948), p. 415.

Schmueser, P., *Feynman-Graphen und Eichtheorien fuer Experimentalphysiker*, Springer, Berlin-Heidelberg, New York, (1988).

When asked why two batteries were used at all, Captain Coler declared that for starting the apparatus a double battery is necessary to get a second charge-impetus after exciting with the one battery, and this for releasing the peculiar character of the apparatus. A test... confirmed this assertion in as much as the mechanism could not be started with the single battery; on the contrary, the "adjustment" of the mechanism got disturbed.

Current-indicators are built into each of the three circuits mentioned, as well as volt-meters, behind some switch resistances necessary for the "adjustment." Between the open ends of the two plate and spool-systems there are the terminal clamps for the effective circuit for whose loading 3 bulbs of 8 volts are provided. The apparatus was then put into action and above all the load was tested with the aid of the built-in instruments, that is on being loaded with [2-3] lamps...

The consumption of energy in the external circuit is greater than the energy taken from the batteries. According to the circuit... the magnet-exciting circuit is fed by a special battery, completely separated from the other two circuits. Consequently, a direct comparison of efficiency and consumption of the apparatus would mean that only the sum of current of the plate circuit and of the spool circuit would count...

The reception of current from the two batteries in this case... was 1.7 watt while the consumption of the bulbs amounted to about 8 watt. Especially striking in this connection is the considerably higher current-power in the bulb-circuit being about 12 times bigger than the current coming from the two batteries...

Prof. W.O. Schumann (Munich) also tested the Stromzeuger in 1926; his six page analysis was included in an appendix of the BIOS Report:

The apparatus in question principally consists of two parallel connected spools, which being bifilarly wound in a special way, are magnetically linked together. One of these spools is composed of copper sheets (the spool is called the plate spool), the other one of a number of thin parallel connected isolated wires (called: spool winding), running parallel at small intervals to the plates. Both spools can be fed by separate batteries; at least two batteries are necessary to put the spools to work.

The spools are arranged in two halves each, according to the bifilar winding system. The batteries are attached to the starting points, and the current-receivers to the parallel connected ends. Inter-communications are connected between parallel windings of the two halves of the plate spool which contain iron rods with silver connections. These rods are magnetized by a special battery through applied windings (called: exciter windings).

According to the statement of the inventor, the production of energy principally takes place in these iron rods, and the winding of the spools plays an important part in it... (The form of the spool is a long small rectangle).

The inventor stated that the apparatus in its installation was very sensitive, especially with regard to the magnetic conditions of the iron cores, and that a wrong treatment [internal measurements] would cause interferences which would be wearisome and very difficult to be eliminated.

The exciter winding is electrically completely separated from the other windings...

Installed in the apparatus were 3 current meters for the currents from the 3 batteries, and furthermore current and volt meters... for the current receivers. One and two bulbs respectively were employed for this purpose...

As a striking fact it should be mentioned that the spool circuit having been at first always switched on alone, received a current of 104 mA. As soon as plates and exciter circuit additionally and simultaneously were turned on, as, according to the inventor, the apparatus demands it, the current in the spool circuit comes down to about 27 mA...

After the present examination, carried through as carefully as [possible], I must surmise that we have to face the exploitation of a new source of energy whose further developments can be of an immense importance... I believe that a further development of the apparatus... will prove justified and of great importance.

In 1943, Hans Coler and Dr. Heinz Frohlich made a report to the Research Department of the Admiralty (OKM) in Berlin, in which they described the Stromzeuger:

The apparatus consists of three principle circuits which are inter-wound and inter-coupled in a peculiar way. Some of these are divided again into single subsidiary circuits which evidently have all to be brought into resonance with one another.

The principal circuits (called the anchor) in which the energy gain probably occurs, consists of metal plates between which transformer coils are connected, the whole being connected to one large coil plate. On each of the single windings on this is coupled a large flat coil (called field). These flat coils are interwound in two groups; these groups represent the turns of a transformer. This couples, on one hand on the flat coils of the other group (as secondary coils), and on the other hand on the anchor-plate coil, which is placed between them. The third electrically independent circuit (called the directing circuit) regulates this transference. Figure 1 [not included in the BIOS Report] shows diagramatically and in plan these interwound parts. Figure 2, the so-called basic diagram, shows the connections between these different parts. This basic diagram shows the conditions necessary for self-interruption. In consequence of this arrangement, different types of currents are created in different conductors (pulsed DC, AC, etc.).

The transformer coils, connected between the anchor plates, are connected in a peculiar way through thin permanent magnet rods. Their main object seems to be to pre-magnetize the transformer cores; it is, however, very probable that they are also connected with the Barkhausen Effect...

In his report Coler also described the following simple, significant experiment:

Given a solenoid consisting of two windings, one upon the other, of the same length and number of turns, enclosing a soft iron cylindrical core. Firmly attached to one end of the core is a pre-magnetized steel rod. If an alternating current is passed through one of these coils, acting as primary, the residual magnetism of the steel rod is strengthened during one half cycle, through the magnetism induced in the core, during the other half it is weakened. If now the other winding is connected in series with the pre-magnetized steel rod (as secondary coil) in such a way that the secondary current must pass through the magnet, one half cycle of the secondary current must be more or less subdued; in other words, a rectifying effect must be created. It may be remembered that according to measurement by Profs. Kloss and Schumann a high-frequency pulsed direct current (about 180 kc) is flowing in the output resistance of the apparatus, for the creation of which no other explanation is possible.

Another experiment with two double-wound solenoids, connected in series, revealed that "in such a secondary circuit a considerable DC component exists which can be strengthened by means which we shall not go into here..." Experiments were conducted with the plate and flat coil unit to test their interactions:

> As the plates are not only charged as condensers, but also have directed currents passing through them it had to be assumed that their mutual influences not only consisted of a condenser effect, but that they also created a magnetic field... It appeared that the "Ankertrakte" AC or BD are not all to be considered as oscillating units, but that the single group systems, consisting of plate - to anchor coils - plate, already represent independent oscillation circuits. *According to this the apparatus contained ten such oscillation circuits*... In order to make the apparatus work the harmony of all oscillating circuits in their individual frequencies would evidently be necessary (at least within certain limits determined by the suppression in question... The previous occasional success must be considered as due to chance.
>
> The mutual influence of the flat coils upon one another could not easily be examined because, despite the great distance between the windings (25 mm), they have remarkably great apparent resistance (about 200 ohms at 10 kc).
>
> It appeared that the power factor of both of the flat coils wound 1:1, in consequence of their peculiar interwinding have the astonishingly high value of 0.85. The value of the power factor was at a maximum at 10 kc, at which frequency the most favorable matching of the impedance... was obtained.

The repeated mention of 10 kc is also found in the scientific literature on the Barkhausen Effect. After all his experiments and measurements, Coler offered the following explanation of the way in which the apparatus works:

> Due to the connecting of the batteries, a current impulse is induced in the anchor circuit which charges the plates. The discharge circuit from the plates causes electrical interruption of short duration of the battery current in the field circuit, which furthermore inductively interrupts, or changes the direction of, the battery current in the directing current for a short time. The electromagnetic field induced by this process in the directing circuit by its dissipation, induces over the field circuit a current in the anchor circuit recharging the plates, and so forth.
>
> Due to the influence of the Barkhausen Effect each single process has an impulse-like character, and the necessary change of phase is produced to allow the regularity of the process.
>
> Due to a source, up to the present not investigated, and not explainable by existing scientific theories, an additional quantity of energy is freed during each cycle which leads to a continual raising of the amplitude of the mutual processes, until the magnetic cores are saturated.
>
> From the fact that in the resistance of the apparatus pulsed direct current is flowing... There is possibly an up to now unknown rectifying effect, or alternatively the gain in energy is produced only during one half of the cycle, either during the charging or discharging of the plates. The activity in the apparatus must take place *in the ten oscillation circuits in a phase-like manner*... No technical means were available to make the necessary tuning adjustments.
>
> It is clear from the above that the success of the inventor up to now could only be due to chance, or happy accident. The necessity, therefore, arises to transfer the apparatus from the state of empirical development, with sufficient technical means and based on results of an exact basic research, to a state of working procedure which can be controlled.

Other experiments, conducted by Frohlich with the arrangement in Figure 6, convinced Coler that his theory was correct:

> The result obtained with this experimental arrangement... was the clear proof of a considerably larger energy during opening (intake), compared with closing impulses... The [10-50%] difference is always in favor of the opening impulse... Consequently this observation can also be considered as a proof of the fact that an energy difference exists... [There is a] considerable, but extremely short, energy peak of the opening impulse... My development of the "Space Energy Receiver" was based on this and was successful.

In the 1980s, George Hathaway (Planetary Association for Clean Energy, PACE) constructed a Magnetstromapparat which produced 50 millivolts, and demonstrated it at unconventional technology conferences. There has been no news of further development.

The Barkhausen Effect upon which the Coler inventions work is a low-field phenomenon that occurs when a ferromagnetic material is subjected to a change in the applied magnetic field. A series of discontinuous steps develop, corresponding to reversals in magnetization domain volumes from 10^{-10} to 10^{-8} cm^3. The size of the discontinuities can be increased by stressing the magnet. Possibly the soft iron magnets which Coler used were subjected to such stress by ultrasonic magnetostriction from the coil windings at the 180 kHz resonant frequency.

One can only wonder why British Intelligence (sic) declassified the Coler report, but we can be glad they did. Would that they had seen fit to include more schematics of the Stromzeuger, because the available diagrams are woefully inadequate. At least we now know a few key facts, to wit: (1) Ferromagnetism resonates at 180 Khz; (2) The Barkhausen Effect can be amplified to provide useful power (10 Khz is another key frequency here); (3) "Some of the magnets are wound in a clockwise direction looking at the N pole (called *left*) and others in an anti-clockwise direction (called *right*)." The resistance of the magnet-coil combination was about 0.33 ohm; (4) "The basic principle is that an electron is to be regarded not only as a negatively charged particle but also as a South magnetic pole;" (5) "The novel feature is that the capacities are connected to the secondary core *through permanent magnets*;" (6) "There is a considerable, but extremely short, energy peak of the opening impulse."

It is to be hoped that the Stromzeuger can be redeveloped, since it is a relatively simple, low technology device requiring no exotic components. It should not be left "to chance, or a happy accident," as it was for Hans Coler. The BIOS Report #1043 is available from Rex Research, P.O. Box 19250, Jean NV 89019 USA for $9 postpaid ($11 overseas).

DONALD P WATSON

RESEARCH INTO ALTERNATIVE TECHNOLGY

12 Chatsworth Road
Bournemouth
Dorset
BH8 8SW
England

Tel:- ***44 1202 302213

ELECTRONICS
RADIO
RADAR
ACOUSTICS
CAVITATION
COLD FUSION
RADIONICS
AETHER SCIENCE
TESLA FIELDS
ENERGY GENERATION
WATER TECHNOLOGY
LEVITATION

Integrity Research Institute
1422 K Street NW, Suite 204
WASHINGTON DC 20005
United States of America

For the attention of:- Thomas Valone

March 1, 1999

Dear Mr.Valone,
I assume that by now you have received my last letter dated 22nd February 1999. Since then, I have received a copy of an Infinite Energy article relating to Hans Coler. I also note that your publications list 1996 features this work.
I am sure that when this data was de-classified there was a deliberate attempt to mystify and confuse because the Infinite Energy article related in the main to practical hands-on experimentation but little to explain a theory of operation.I was almost on the point of trashing this paper when I looked again at the first three diagrams Issue 21,page 30.
Instant recall to 1960 and a course on marine wireless Telegraphy brought to mind a 100Hz bandpass filter on the Marconi Atalanta Double conversion Superhet which functioned as a Magneto-strictive filter. Just like Colers diagram , a rod of either ferrite or magnetic material had a coil at both ends, the driven coil was screened to prevent stray coupling and when energised caused the rod to vibrate longitudinally creating a vibrating magnetic field which cuts the windings of the second coil functioning as a generator.The difference being is that the pick-up coil generates a frequency of twice the input and this can be counteracted by a third coil carrying a direct current to polarise the core material. What continues is just me thinking out loud but, which may have some bearing on resolving the modus operandi.
In the Coler diagram there are only two coils which seems to suggest that if all six magnets are mechanically connected as in a torus, each stage creates a voltage doubling effect.
We should also ask, why the two coils on each magnet are wound clockwise for one and anti-clockwise for the other. This would seem to suggest either a polarising effect on the core or alternatively a standing wave effect on each segment such that the vectors cancel and a scalar wave is generated
Also,since all six units are mechanically coupled the unit should create an increasing frequency and positive feedback with gain, the requirements for any type of oscillatory circuit.If no visible input of energy is evident then perhaps we have a scalar amplifier/converter to electromagnetics or perhaps the reverse is the case. I am not sure where the copper plates fit into this unless rather than behaving as a capacitor they operate as an energy pump between stages as was theorised in the functioning of the Grasshopper project.

There does seem to be some similarity with the Kleinian Roll Oscillator by Takamura featured in Principles of Ultrarelativity, regrettably my maths standard does not permit an in depth study. An investigation of scaler generators and detectors by T.E.Beardon in Fer de Lance may also be helpful and for a detailed study of Magneto-striction I suggest chapter 4 of Ultrasonics by Gooberman,English Universities Press
To conclude,circa 1940, a Prof Maby was using scalar radiations as a primitive radiolocation device prior to the invention of the Magnetron.Of significance is his statement that all elements radiate scalar waves which rotate through 360 degrees every 24 hours,which can be detected with a simple neon device and that the wavelength can be calculated in centimetres by the square root of the atomic number times the atomic weight.

Since my brain gets a bit addled with old age perhaps someone can take this discourse a little further.I have not attempted an experimental model and therefor I am not able to comment further. But readers who have an interest may wish to advise me how near the truth my theories are.

Sincerely

Don Walton

Off the Shelf...

Space Energy Receivers

by Simplified Technology Service

Reviewed by: Michael A. Tyborski
Systech Electronics
10953 West Mitchell Street
Milwaukee, WI 53214

Important inventions tend to slowly gain acceptance or get lost in controversy. Energy converters are no exception. Since they rely on an unknown energy source, these devices engender ridicule from orthodox scientists. Never-the-less, they continue to intrigue lone inventors.

As an unimpressive looking booklet, Space Energy Receivers may be a gem in disguise. It describes early devices that tapped this hidden energy source. These converters used coils, magnets, and common materials to produce useable electric power. This hardware, although crude by today's standards, shows the validity of ether technology.

Most ether research was done before 1920. This booklet, however, concentrates on devices that were publicly demonstrated before 1945. By using obscure news clippings and patents, it describes inventions that are not mentioned elsewhere. Throughout its twenty-one page text, references are listed for further study.

The author begins by describing early attempts to prove the ether's existence. As the basis for all matter/energy, this quest has proved to be very difficult. Evidently, physicists fail to see that we exist within the ether like fish in the ocean. The famous Michelson-Morley experiment disproved the ether on false grounds! Later experiments by Silvertooth and Sagnac, fortunately recognized the problem and compensated for it.

The material shows that some factors are important. First, collector assemblies must be isolated from ground. This applies whether plates or wires are used. One design even used carbon rods in a circular pattern. Another, the "Perrigo", used plates to obtain 373 kilowatts per square foot of surface area. It was shown to the 65th U.S. Congress in 1917.

Second, magnets should be used as active converter elements. Roy Meyers used specially magnetized steel plates in his collector. Other converters used magnets within transformer cores. Actual designs, unfortunately, were omitted.

Finally, successful designs must compensate for power variations. Most inventors saw changes based on solar activity and the time of day. In one case, drafts were significant!

Nikola Tesla appears to have created this field. He stated:

> *...ere many generations pass our machinery will be driven by powers obtainable at any point in the universe . . .*

In 1901, he obtained a patent for his "Method of Utilizing Radiant Energy." Throughout the next forty years, he continued to show that he knew much more about this energy source.

Tesla believed that electromagnetic fields travel longitudinally through the ether like sound waves. He even argued with Heinrich Hertz over the Hertzian wave theory that relied on transverse motion.

During this period, Henry Moray developed his converter. This elaborate unit could produce up to 50,000 watts of power. Surprisingly, the entire assembly only weighed 50 pounds! Moray closely followed Tesla's work and considered space to be a "Sea of Energy."

The final section introduced Hans Coler's "Stromerzeuger." This device could produce 6 kilowatts from a dry cell. The British government, following the German lead, became interested and interviewed Coler and his associates. Their final report included evaluations from other scientists.

The author reproduced some pages from British Intelligence BIOS report 1043. Since its declassification, the Imperial War Museum in London provides copies of the complete document.

Coler found that magnets oscillate at about 180 kilohertz. He also noticed that flux line swirl varies at each pole. As a result, he wound coils in different directions depending on the polarity of closest magnetic pole. His research notes, however, did not explain how the converter worked.

Space Energy Receivers, although brief, provides an important introduction to free energy systems. It offers inspiration for experimenters working with common components. And more important, it lays a foundation for advanced research.__*MAT*

FOR FURTHER INFORMATION ON BIOPHOTONS AND BIOELECTROMAGNETICS ORDER THESE BOOKS AND REPORTS

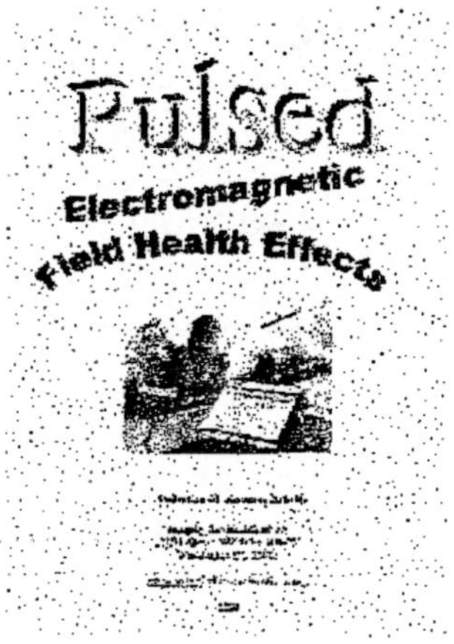

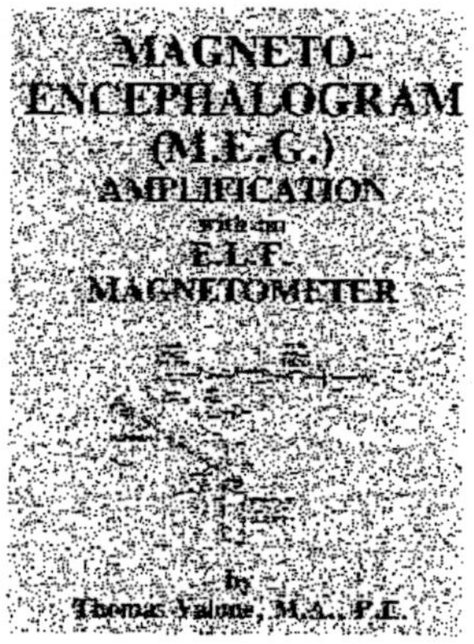

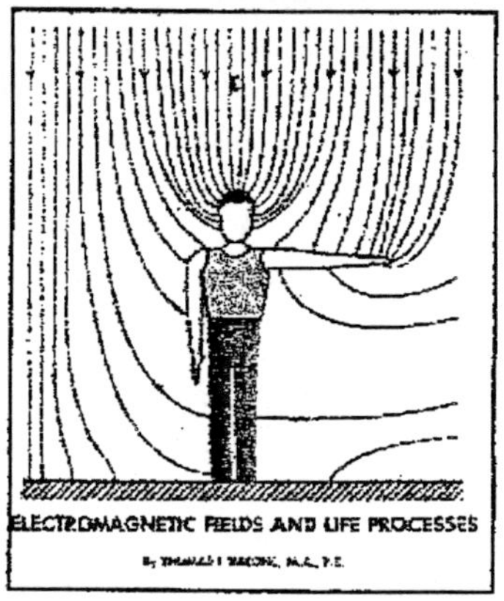

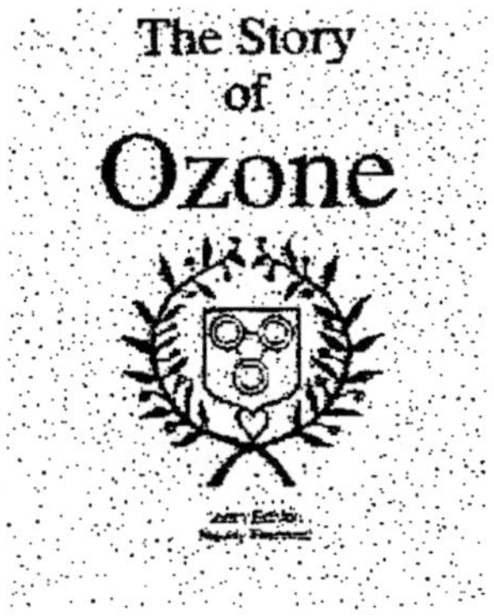

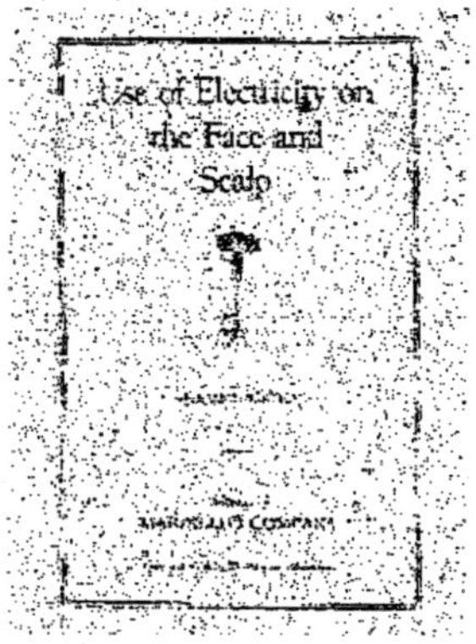

ALL AVAILABLE AT :
www.IntegrityResearchInstitute.org.
Click on order page

Tape Here

Thank You

If you wish to receive a FREE copy of our latest Future Energy Publication catalog and to be placed on our annual mailing list, please cut out, fold and mail back.

Where did you purchase this book? ____________________

Email for Future Energy monthly: ____________________

Name ____________________

Address ____________________

City ____________________

State ________ Zip ____________ Country ____________

← *Fold Here* →

Non-Profit Institute Dedicated to Scientific Integrity in Emerging Energy Research

PLEASE PLACE STAMP HERE

Integrity Research Institute
5020 Sunnyside Avenue, Suite 209
Beltsville MD 20705

978-1-935023-06-7
9781935023067